Railway Vehicles and Infrastructure

Railway Vehicles and Infrastructure

Guest Editors

Pedro Antunes
Hugo Magalhães
Pedro Aires Montenegro

MDPI

Basel • Beijing • Wuhan • Barcelona • Belgrade • Novi Sad • Cluj • Manchester

Guest Editors

Pedro Antunes
Institute of Railway Research
University of Huddersfield
Huddersfield
UK

Hugo Magalhães
Instituto Superior Técnico
Universidade de Lisboa
Lisbon
Portugal

Pedro Aires Montenegro
Faculty of Engineering
University of Porto
Porto
Portugal

Editorial Office

MDPI AG
Grosspeteranlage 5
4052 Basel, Switzerland

This is a reprint of the Special Issue, published open access by the journal *Vehicles* (ISSN 2624-8921), freely accessible at: https://www.mdpi.com/journal/vehicles/special_issues/GW15F1ESL9.

For citation purposes, cite each article independently as indicated on the article page online and as indicated below:

Lastname, A.A.; Lastname, B.B. Article Title. *Journal Name* **Year**, *Volume Number*, Page Range.

ISBN 978-3-7258-7270-1 (Hbk)
ISBN 978-3-7258-7271-8 (PDF)
https://doi.org/10.3390/books978-3-7258-7271-8

Cover image courtesy of Hugo Miguel Pacheco Magalhães

Contents

Preface

Railway vehicles play a key role in the global transportation sector. Whether moving passengers or freight, they operate under strict performance, safety, and efficiency requirements while continuously interacting with an increasingly complex rail infrastructure. The growing demand for cost-efficient, environmentally sustainable, and fast mobility solutions places continuous pressure on research and development efforts, further driving innovation across the railway domain and encouraging the exploration of increasingly sophisticated engineering approaches.

Research in railway systems is inherently multidisciplinary. A thorough understanding and accurate prediction of the coupled behavior between the vehicle and the infrastructure, encompassing the track, foundations, bridges, tunnels, and the overhead line system, remains essential. Recent progress in numerical modeling has enabled the simulation of highly detailed dynamic effects, providing accurate and reliable analyses that support both scientific discovery and industrial applications. In the railway context, such tools contribute not only to improving our understanding of system behavior but also to enhancing design processes, supporting optimization, and informing homologation procedures. As a result, they help accelerate the development and validation of new railway technologies. Experimental testing continues to play a crucial role, and its combination with virtual testing environments increasingly fosters a richer and more comprehensive framework that strengthens research robustness and accelerates the emergence of new findings.

This Special Issue aimed to bring together high-quality contributions that advance knowledge on railway vehicles and their interaction with infrastructure. The articles presented in this collection illustrate the diversity and depth of current research activities in the field. Together, they highlight emerging methodologies, innovative technological solutions, and rigorous scientific approaches that contribute to improving the safety, reliability, and performance of modern railway systems. We hope that this Special Issue will serve as a valuable resource for researchers, engineers, and professionals engaged in the development, analysis, and long-term understanding of advanced railway technologies.

We would like to thank all authors for their valuable contributions and the reviewers for their careful and constructive evaluations, which helped ensure the high scientific quality of this Special Issue. We also extend our gratitude to the editorial team of Vehicles for their continued support throughout the preparation of this issue.

Pedro Antunes, Hugo Magalhães, and Pedro Aires Montenegro
Guest Editors

vehicles

MDPI

Article

Investigation of the Causes of Railway Track Gauge Narrowing

Péter Bocz [1], Nándor Liegner [1], Ákos Vinkó [1] and Szabolcs Fischer [2,*]

[1] Department of Highway and Railway Engineering, Faculty of Civil Engineering, Budapest University of Technology and Economics, 1111 Budapest, Hungary; bocz.peter@emk.bme.hu (P.B.);liegner.nandor@emk.bme.hu (N.L.); vinko.akos@emk.bme.hu (Á.V.)
[2] Department of Transport Infrastructure and Water Resources Engineering, Faculty of Architecture, Civil and Transport Engineering, Széchenyi István University, 9026 Győr, Hungary
* Correspondence: fischersz@sze.hu; Tel.: +36-(96)-613-544

Abstract: On behalf of MÁV Hungarian State Railways Ltd., the authors carried out a research and development (R&D) project on behalf of the Budapest University of Technology and Economics, Department of Highway and Railway Engineering, on the subject of "Research and investigation of the causes of gauge narrowing by finite-element modeling in running track and turnout, and under operational and laboratory conditions". The main objective of the research was to investigate the causes of localized defects of gauge narrowing in railway tracks based on machine and manual track measurements, laboratory measurements, and theoretical considerations. The measures proposed as a consequence of identifying the causes could significantly contribute to reducing the number and extent of local defects in the future. Furthermore, the research aims to develop new theories in less scientifically mature areas and provide procedures and instructions that professional engineers and practitioners can easily apply. The main areas of research, which are not exhaustive, are as follows: (i) the evaluation of the measurement results provided by track geometry measuring and recording cars; (ii) on-site investigations in the railway track in terms of gauge and rail profile measurements; and, based on these, (iii) the selection of concrete sleepers, which were removed from the track and subjected to more detailed geometrical investigations in the laboratory, together with the components of the rail reinforcement; (iv) the track–vehicle connection, tight running in straight and curved track sections under track confinement; (v) modeling of the stability and deflection of the rail when the rail fastenings lose part of their supporting function; and (vi) finite element modeling of the concrete sleepers under operating conditions such as slow deformation of the concrete, temperature variation effects, and lateral support on the ballast. In the already-narrowed track section, the tight vehicle running is not the cause of the track gauge narrowing but a consequence, so it is not investigated in this paper.

Keywords: railway track; track gauge; track gauge narrowing; plain track; turnout; statistical analysis; finite element modeling

1. Introduction

1.1. General Introduction

Transport includes land, sea, and air transport [1–3]. Air and space transit should be combined in the discussion of the topic.

Transportation has always been an important part of human society, allowing people, products, and services to move from one location to another. From early human migration to the modern global economy, the necessity for transportation has developed throughout time.

Walking was the original mode of ground transportation, utilized for thousands of years. Later, humans tamed animals such as horses, camels, and oxen to aid in transporting goods and people. Carts and chariots were developed with the invention of the wheel in ancient Mesopotamia, leading to the earliest road networks.

The 18th and 19th centuries saw considerable developments in transportation as a result of the Industrial Revolution. The invention of steam-powered vehicles, such as trains and steamboats, transformed transportation and commerce.

The Wright brothers' aircraft made the first successful powered flight in 1903. Air travel grew increasingly popular in the twentieth century, and it is now the quickest option to travel great distances. As a result, air travel has significantly influenced worldwide commerce, tourism, and globalization.

The first satellite was sent into orbit in 1957; space transportation is a relatively new invention. Since then, people's flights to the moon and spacecraft launched to study other planets have extended space exploration. As a result, space travel has been critical in scientific study, the military, and communication.

For thousands of years, boats and ships have been used for fishing, commerce, and warfare. However, the invention of steam-powered ships in the nineteenth century changed ocean and river transportation. Ships are now the principal form of transportation for bulk commodities like oil and grain.

Land transport since antiquity requires roads, ports, and docks [4–8]. Modern innovations also necessitated trains and airports [1–3,9–12]. Logistics, along with transport, became crucial [13–17].

The current article discusses the details of railway track gauges, i.e., the track gauge narrowing. The evolution of fixed-rail transportation is beneficial to introduce the issue and establish its topicality. In the mining business, wagon tracks were the first fixed-rail tracks. After "horse-drawn locomotives," George Stephenson invented the steam locomotive. In 1825, the Stockton–Darlington became the first steam railway. Steam locomotives transformed rail transit.

George Stephenson's Rocket, the "official" first steam locomotive, had a track gauge of 1422 mm (4 feet 8 inches) and was eventually extended by 13 mm. It favored wheel flange lateral forces [1].

It may have been the Roman road vehicle track gauge, according to Pompeii and other discoveries. Everything was measured then. Two persons might easily sit on each side of a corridor at 1435 mm [1].

The gauge value has covered 1435 mm, the nominal track gauge, for decades.

In cross-section, the track gauge is the shortest distance between the rail heads' inner guiding edges in the plane parallel to the rail heads' top tangent. The considered depth below the rail head top tangent (Z_P) is 14 mm for Vignol rails and 9 or 10 mm for grooved rails based on EN 13848-1 [18].

To be able to achieve the aims of this article, research was conducted into a selection of the global literature on the topic of railroad track gauges. This research was conducted globally, although it could not be thorough or complete due to space constraints. However, an attempt was made to summarize the content deemed most significant and referenced the most often; the results of this endeavor are summarized in the following paragraphs. The authors gathered the papers and organized them according to the following primary topics: (i) the development of track gauges around the world, as well as their distribution and connecting history; (ii) the measurement possibilities of track gauge(s) (it is important to note that only the non-commercial and non-standardized methods are discussed and detailed in the Introduction; the standardized methods are partly discussed in Section 2); (iii) the degradation of track gauges over time; and (iv) the degradation of railway tracks.

1.2. Literature Review

In the following paragraphs, the authors will depict the first group (i) mentioned in Section 1.1. This organization is interested in the evolution of track gauges across the globe, as well as the history of their distribution and relationship.

Czére [1], as well as Chandra and Agarwal [19] discussed the evolution of track gauges in the past in a very detailed manner. The gauge of a railway track considerably influences a railway system's efficiency and safety. Railway track gauge development has been an extended process covering many decades, with several stages of standardization.

Tzanakakis [20] presents a track gauge creation hypothesis. He begins with Bronze Age vehicles, Roman Empire mythology, and Julius Caesar's traditions. Next, he measures longitudinal tracks and road vehicles' "wheelsets." Finally, British engineers and George Stephenson's locomotive established the standard gauge of 1435 mm and English railroads of comparable construction in the 1800s.

Monfort [21] examines Spanish standard gauge railway construction from 1989 through 2019.

Puffert [22,23] studies track gauge standardization. His book [22] covers North American railways between 1830 and 1890. In the 1830s, several track gauges were used in the US, Canada, and elsewhere. Nine gauge-valued areas were formed by 1860.

Wang et al. [24] examined track gauge dispersion across Eurasia. Eurasia has 11 railway track gauge values: broad, standard, and narrow. 1520, 1435, and 1067 mm are typical. Track gauge coverage and geographical dispersion vary greatly. These may divide rail regimes and networks. Eight rail geosystems resulted, with 1435 and 1520 mm gauge gaps being enormous. China and neighboring nations are quickly building standard gauge railroads. Standard gauge high-speed railroads have recently connected Moscow–Beijing and Beijing–Singapore. However, high-speed trains can only partly link neighboring nations with various track gauges. Technology, route dependence, geopolitics, national security, and military-width colonial terrain impact Eurasia's rail geo-system, which has five geo-modes of rail gauge spreading.

The aforementioned (ii) group (see Section 1.1)—track gauge measurement possibilities—is represented in the following paragraphs. (It must be mentioned that only the non-commercial and non-standardized methods are overviewed and detailed here in the Introduction; the standardized methods have been dealt with in Section 2).

Zheng et al. [25] devised a portable trolley system for evaluating railway track gauge and rail deterioration. The approach uses a laser vision inspection model and system calibration. The method involves three-dimensional machine vision.

Pay et al. [26] devised and built an IoT-DTG (Internet-of-Things-enhanced digital track gauge) to measure track gauge and superelevation. (in other words: cant or cross-level). It has a better measuring capacity and accuracy than manual track gauge measurement devices. The innovative technology saves 50% labor, 60% efficiency, and SGD 19,200 annually (Singapore dollars).

Yilmazer et al. [27] created a drone-based optical and image analysis track gauge measurement system. The drone must fly at a certain height, unaffected by the cars. Image analysis removes vibrations. Gaussian filters and unique detection techniques may identify tracks in photos.

Zhang et al. [28] used wheel and rail transverse motion to estimate track gauge. Two laser sources and two cameras dynamically capture the rail head's inner side. Rail profile dictates parameters. The track gauge is computed using the wheel's lateral displacement and the rail head's laser point's vertical displacement. The experimental findings demonstrate the system's simple hardware architecture, little data calculation, excellent measurement accuracy, and non-contact gauge-spacing measuring.

Shi et al. [29] designed a track-detecting trolley-mounted laser triangulation tracking system. Two 2-D laser scanners collected left- and right-track profile data. An iterative closest-point (ICP) method calibrated sensor data, and an adaptive filter technique smoothed the contour. To be able to increase positioning accuracy, the track gauge was fitted to the rail head curve using least squares. According to studies, the suggested approach enhances track gauge measuring properly and effectively with a ±1 mm precision.

Tsubokawa and Ishikawa [30] designed railway track measurement equipment that can be installed on a vehicle that measures gauge, twist, and loaded track conditions. Laser-based, non-contact measuring technology "scans" rails in two dimensions at a track cross-section. The maximum measurement speed is 40 km/h. A motor automobile was measured many times across a 400 m radius curve with a 30 mm superelevation. Four sleepers had their rail fastenings removed to test railway gauge variance. A standard railway wagon was installed to measure a 120 m site radius track segment with 107 mm superelevation and 1067 mm gauge at 10–40 km/h several times. Track gauge repeatability was 0.3 mm, and 5 m base length twist repeatability was 1.0 mm.

Tang et al. [31] used a 2D laser scanner to measure grooved tramway rails. It measures +/−0.6 mm accuracy with the "laser triangulation" technique. Measured rail head profiles were first mathematically filtered. Then, the rail head detection and gauge measurement were lab-tested.

The authors discuss the aforementioned (iii) and (iv) groups (see Section 1.1)—track gauge degradation in fixed-rail applications (tramways, railroads, etc.) and track gauge-specific regions and results—in the following paragraphs.

Németh et al. [12] examined track gauge changes over time on five low-traffic and five high-traffic Hungarian railway lines with straight track portions. Calculations included through-rolled axle tons. (i.e., the track loading, MGT—million gross tons). The measurement data were "produced" by the FMK-007 track geometry measuring vehicle owned and operated by MÁV CRTI Ltd. (MÁV Central Rail and Track Inspection Ltd., Budapest, Hungary). Twenty-five cm track geometry was measured raw. Data analysis included mean, standard deviation, skewness, kurtosis, and Vaszary-type shape numbers of distribution functions. The eleven railway lines were also spectral analyzed. The track gauge change, average and standard deviation readings, elapsed time, and traffic loads did not correlate. (i.e., MGT). The gauge value distribution functions for the ten lines were neither Gaussian nor lognormal. Distribution functions are flatter at lower track speeds and steeper and narrower at higher speeds.

Fischer et al. [10] investigated the change in track gauges in curves for numerous railway lines in Hungary with low and heavy traffic (the same railway lines as in [12]). The curves included transition curves (TCs), circular curves (CCs), and entire curves. (ECs). The distribution functions were used to assess the change in the track gauge parameter as a function of elapsed time. The Kolmogorov–Smirnov test was used to conduct a statistical test, and the skewness and kurtosis parameters were computed and examined. The scientists determined that the kind of track system and sleeper type also had an impact on the change. The publication includes the precise deterioration ratios for every case study.

Track inspection train field data are used to analyze track geometry abnormalities' time-frequency characteristics [11]. The cumulative difference from the mean value is obtained using auto-adaptive signal decomposition on four pre-selected track sections. The sensitive wavelengths of track abnormalities are acquired from the suitable allocation of wavelength ranges in the Fourier Amplitude Spectrum (FAS) of the original signal and the Fourier transform of the components identified by the Variational Mode Decomposition (VMD). Track Quality Index (TQI) can detect track faults and right/left rail waveforms. The track gauge is broad-band, whereas the alignment parameter is narrow-band. The "manual" component separation and auto-adaptive component decomposition approaches were examined independently. Empirical mode decomposition's (EMD) original track gauge Intrinsic Mode Function (IMF) was non-stationary. Variational Mode Decomposition separated the clustered components within the IMF retrieved by EMD and the physically relevant trend of the parameters, producing equivalent results to FAS-based manual component separation. The extracted modes' spectral peaks are thought to come from track-vehicle mutual excitation and need vehicle dynamics calculations. This extra information regarding rail irregularity waveforms will help railway engineers identify early track stability issues and investigate track anomalies.

Aharkov et al. [32] examined the gauge stability of KB-75, KKP-1, and KPP-5 rail fasteners utilized in the Ukrainian railway network. Mathematical statistical ap-proaches compared railway track measurements. The study used 12 data sets from three-monthly assessments over three years. The study's straight line has 60 million MGT of mixed passenger and freight train traffic.

Ahac and Lakusic [33] modeled tramway gauge deterioration for maintenance planning. They created a theoretical-practical mechanistic-empirical deterioration model. They studied track gauge adjustments (primarily widening) on Zagreb's 1000 mm gauge tramway network. They examined Ri-60 "grooved rail" rail profiles and R200 and R260 rail grades for R > 200 m and R < 200 m site radii. Tracked and direct rail reinforcing were proposed. This study examined a 4.5 km portion of 25 Zagreb tramway segments constructed and rebuilt between 1997 and 2004. Track geometry (only straight sections were considered), traffic characteristics (only "open track" sections with separated track structure, no tram stops or level crossings, and a constant average speed of 15 km/h were considered), and rail quality and rail reinforcement were homogeneous.

Akkermann and Akkermann [34] addressed rail gauge trajectory and track micro- and macro-profile. In addition, they presented a track gauge parameter-monitoring digital clone in their article.

Falamarzi et al. [35] used an ANN (artificial neural network) and SVM (support vector regression) to forecast electrified track gauge variation. Each model's goodness of fit was determined using R^2 and MSE (mean squared error). The Melbourne light rail network was studied: it is a 250 km double-track segment with 25 lines. In 2013 and 2014, non-contact optical measuring equipment measured a 10 m string length. Yarra Trams supplied statistics. First, mathematical filters removed outliers and noise from measurement data. Both ANN and SVM techniques yielded excellent results, although the deterioration model based on the ANN method predicted straight track portions more accurately than curved track sections.

1.3. Novelty of the Current Paper

Based on the above literature review, the authors dealt with analyzing the track gauge parameter, i.e., one of the most critical geometrical characteristics of railway tracks.

The authors conducted a research and development (R&D) project for MÁV Hungarian State Railways Ltd. on behalf of the Budapest University of Technology and Economics, Department of Highway and Railway Engineering. The project was entitled "Research and investigation of the causes of gauge narrowing by finite-element modeling in running track and turnout, and under operational and laboratory conditions" (see also the last paragraph of the Section 1.1). Using machine and manual track measurements, laboratory measurements, and theoretical considerations, the study's primary purpose was to investigate the causes of localized track gauge narrowing defects in railway tracks. The solutions proposed as a result of identifying the causes have the potential to reduce the frequency and severity of future local faults substantially. In addition, the study aimed to establish new concepts in less scientifically mature fields and provide techniques and instructions that expert engineers and practitioners can readily implement. The tight vehicle running is not the cause of the track gauge narrowing but a consequence, so it is not investigated in this paper.

2. Evaluation of the Track Gauge Data and Exploration of Possible Correlations between Error Points and the Superstructure System, Track Geometry, and Loading

2.1. Measurements with Measuring Cars

The MÁV Central Track Inspection Ltd. (MÁV KFV Kft., Budapest, Hungary) performs track diagnostic measurements on the network operated by MÁV Zrt. (Hungarian State Railways Ltd., Budapest, Hungary) on the order of MÁV Zrt. two or three times a year. TEN-T (Trans-European Transport Network) lines are monitored with the FMK-007 measuring car and other lines with the FMK-004 measuring vehicle. The measuring car FMK-007 can measure at a speed of 200 km/h and is equipped with two measuring systems: a track geometry measuring system and a vehicle dynamic measuring system. The geometric measuring system provides distortion-free and contact-free data using optical and inertial measuring systems. The FMK-004 is a Plasser EM-120 self-propelled rail vehicle capable of measuring at 100 km/h. The measurement system is contact-based, i.e., the instruments on the central bogie mounted on the vehicle and the measurement sensors mounted on the front and rear bogies by contact with the track. The measured raw data are stored at 0.25 m and evaluated as follows for both measuring cars:

- Track measurement graph showing the values of the measured characteristics by chainage;
- List of local geometric errors for the measured features;
- General track condition rating in the form of a rating number;
- Statistical evaluation.

The track measurement graph displays the geometric data on a graph similar to a long-section, which can be generated by loading the raw data into an office program and exported to PDF.

To be able to produce a list of local defects, it is necessary to have a set of dimensional limits against which the "defective" sections can be evaluated. MÁV Zrt. uses a multi-step size limit specification. For the purpose of this research, the measurement data and the size limits are required for the track gauge; the specifications of MÁV Zrt provide the size limit values. D.54, Chapter 51, is summarized in Table 1 [36]. The negative value of the track gauge (gauge narrowing) does not include a "D" category. Therefore, when assessing the gauge, not only local errors but also the moving average of the gauge values over 100 m shall be considered.

Table 1. Limit values of track gauge according to specifications of D.54 [36].

Speed	Gauge Widening [mm]							Gauge Narrowing [mm]					
[km/h]	A1	A2	B	C1	C2	C3	D	A1	A2	B	C1	C2	C3
0–80	+5	+10	+15	+25	+30	+35	+45	–4	–4	–5	–7	–8	–9
81–120	+5	+10	+13	+20	+25	+30	+45	–4	–4	–5	–7	–8	–9
121–160	+4	+7	+11	+15	+20	+25	+45	–4	–4	–5	–6	–7	–8
161–200	+3	+5	+9	+10	+15	+20	+45	–3	–3	–4	–5	–6	–7

The following tolerances are determined and detailed in [36].

A1 Construction limit: The gauge category whose limits shall not be exceeded by the geometric dimensions of a track constructed of new material (rail, sleeper, fastenings, new or renewed ballast in whole cross-section) on a new or renewed substructure.

A2 Construction limit: A gauge category whose limits shall not be exceeded by the geometric dimensions of the track constructed of used material (rail and/or sleeper not new and/or ballast not new or renewed).

B Maintenance limit: The gauge category whose limits shall not be exceeded by values achieved after maintenance works.

C1 action limit: Warning size limit category, which indicates the progressive deterioration of the track geometry. The evolution of the number of defects shall be monitored by comparing current and previous measurements.

C2 action limit: Intervention size limit category, which, if exceeded, requires maintenance work if the evaluation indicates that the local defect size is expected to reach the limit of C3 by the next track inspection measurement.

C3 action limit: Immediate action size threshold category, which requires action to be taken if reached: introduction of a speed restriction within a maximum of three days. After that, maintenance works of the track geometry must be carried out to remove the speed limit.

D closure limit: a limit category independent of the speed above which the track must be closed.

2.2. *Evaluation of Local Gauge-Narrowing Errors*

The local gauge-narrowing data were available in Excel format based on measurements taken in the first half of 2015. In addition, the location data of the local faults were used to determine the type of superstructure (in terms of the rail system and superstructure type) at a given location.

The local faults were plotted on a diagram using proprietary software, as shown in Figure 1. Stations and chainages were also displayed for easy location identification. The diagrams are mainly used to determine the number of defects and the distribution of defects. The height of the bars in the graphs is proportional to the value of defects, but values below the C1 size limit are not shown here as they are not considered measure size limits and are not included in the processed local error data set. It is, therefore, essential to emphasize that the dense columns indicate that the local values of gauge narrowing are densely above the local C1 error limit, but the sparsity of the columns does not mean that the gauge narrowing values on the given track section may not be close to the C1 size limit. As in Figure 1, the entire railway network has been evaluated for local gauge errors.

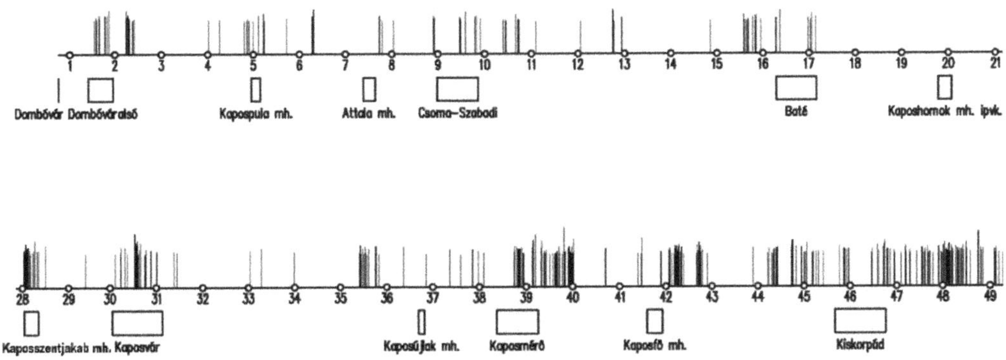

Figure 1. Example of diagram indicating local gauge-narrowing errors ("mh." means station stop, "ipvk." means siding connection).

2.3. Relationship between the Location of Track Sections with Gauge Narrowing and the Type of Superstructure

The study was carried out on the TEN-T network using the results of the FMK-007 measuring car. The Hungarian railway network's most commonly used rail systems are MÁV 48.3, MÁV 48.5, 54E1, and 60E1. The study determined the total lengths of local defects by superstructure type based on the measurements of the MÁV track measuring car in the first half of 2015. The results show that even the railway lines with a rail system of 60 kg/m and almost exclusively with Skl-1 or Skl-14 tension clamp rail fastening only marginally reach the C1 gauge limit. It is due to the fact that these are the youngest sections on the national network.

Failures are more common in the tracks with 54 kg/m rails, but the frequency of mass occurrence is higher in the older 48 rail-system structures, which contain wooden inserts in the sleepers for the bolts of the fastenings.

Although the 54 rail superstructure also has a significant length of gauge narrowing, the length of tracks with this superstructure is also greater than that of the 48 kg/m system tracks. The results show that while 0.15% of the 54-system superstructure has at least C1 gauge narrowing, the same value for the 48.3 and 48.5 superstructures is between 0.79 and 1.34%, which is significant. Sections exceeding the C3 size limit are around 6% of the total length of sections exceeding the C1 size limit.

2.4. Relationship between the Location of Track Sections with Gauge Narrowing and the Alignment of the Track

The correlation between local gauge errors and curvature was investigated for the TEN-T network using the results of the FMK-007 measuring vehicle. Table 2 shows the local errors of the straight and curved sections in percentage distribution. It can be seen that the gauge errors are concentrated in straight sections, with very few errors in curves.

Table 2. The relationship between curvature and gauge errors.

Curvature of Track	Distribution of Local Errors in %		
	C1	C2	C3
Straight	95.5%	94.5%	93.3%
Curve	4.5%	5.5%	6.7%
Sum	100.0%	100.0%	100.0%

2.5. Detailed Analysis of Selected Railway Lines

For more in-depth studies, detailed raw data, measured at every 0.25 m, were also obtained for five main railway lines over several years, one of which is the 100a Budapest-Nyugati–Cegléd–Szolnok railway line. The analyses were carried out for the first half of 2010, 2013, and 2015 to identify the sections where track gauge C1 has not yet developed but is already present to some extent and the rate to which the gauge changes over time.

In the first step, the gauge values were plotted on a graph. The graph is similar to the track measurement graph but contains only the track gauge values in different colors, superimposed for the three years. According to the raw data format, the values are plotted with a deviation from the nominal 1435 mm.

The track measurement graph clearly shows the known correlations, such as the wear of the outer rails in the curved sections and its progression as a function of time. In addition, the sections with different gauge averages are distinguishable.

However, the different stages were not determined by visual inspection but by the so-called cumulative deviation from the mean method. The steps of the cumulative deviation from the mean method are as follows:

- Determine the average of the individual values of the track gauge (TG) of the line in the test ($\overline{x}_{TG}$):

$$\overline{x}_{TG} = \frac{\sum_{i=1}^{n} x_{TG,i}}{n} \quad (1)$$

- Calculation of the value of the cumulative deviation (S_{TG}) from the mean at a given measurement point '*i*': its value is the sum of the deviations from the mean of all points between point 1 and point *i*,

$$S_{TG,i} = \sum_{i=1}^{n} (x_{TG,i} - \overline{x}_{TG}) \quad (2)$$

- Display the values on a graph (Figure 2).

Different attributes have been assigned to the sections—such as stations, type of superstructure, and curvature—to help to identify the location and draw conclusions.

For five selected railway lines, the division into homogeneous sections and the time series averages for each section were analyzed line by line, and the conclusions were drawn.

2.6. Analysis of the 100a Budapest Nyugati-Cegléd-Szolnok Line, Right Track

To analyze gauge variation, the curved and straight track sections have been separated. The remaining sections were classified as "mixed" sections, with several curves and intermediate straight lines.

For purely curved sections (or sections with several consecutive curves), the overall change in gauge over time is that there is no contraction in these sections, but the gauge increases due to the wear of the outer rail. Based on this, the rail's gauge expansion and wear rate were found to be 0.47–0.50 mm/year.

The variation of the gauge in straight sections with time is shown in Figure 3. Many longer straight sections were investigated along the entire line, 20 in total. These straight sections are shown in different colors, with the initial chainage of the straight section on the legend. The rate of change of gauge is slightly variable for each section, based on the individual graphs, with the 2013 measurement showing an increase of 0.5–1 mm compared to the previous year, followed by a decrease of 0.5–1 mm in 2015. The regression line plotted on the mean of the values is marked with a bold red dashed line.

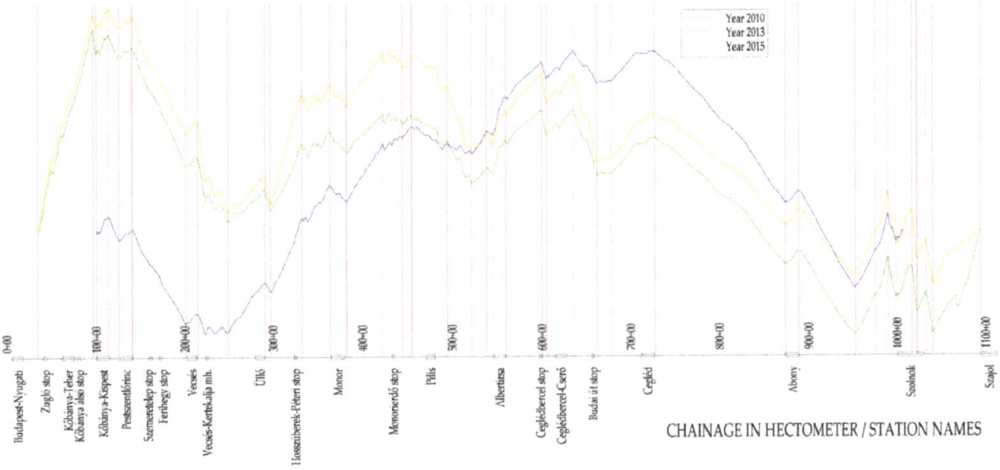

Figure 2. Example of dividing the track into sections by the cumulative deviation from the mean method (the red vertical lines illustrate the considered section borders for the calculation of cumulative deviation between Budapest-Nyugati and Szajol railway stations).

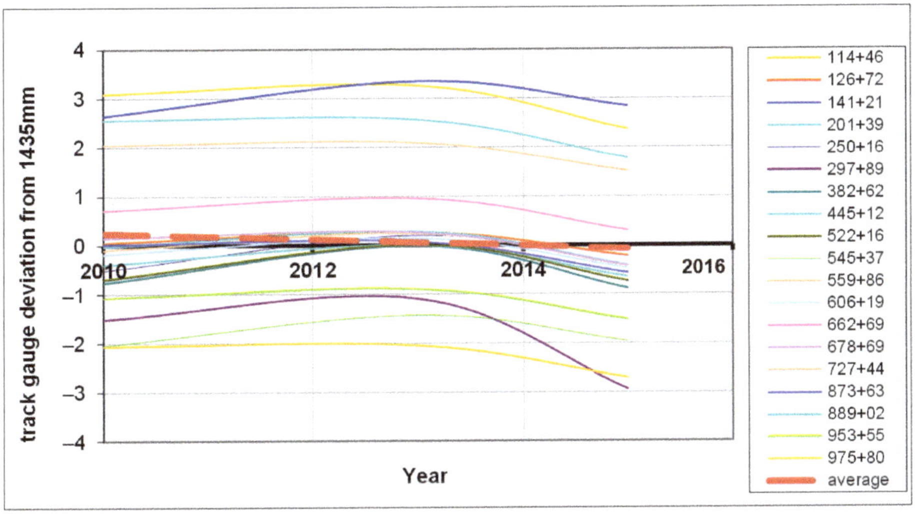

Figure 3. Change of gauge in straight sections of line 100a in the function of time.

The curves shown in the graph show a wide variation, possibly due to the track structures' different age, type, and conditions. It is, therefore, worth grouping the curves according to the superstructure system, as shown in Figure 4. Figure 4 shows that sections with 54 rail systems typically have a small gauge narrowing (1.0–2.5 mm), while sections with 60 systems tend to have a normal value or a small narrowing but with a considerable variation. The sections that differ in the "upstream" direction, i.e., in the direction of the widening, are the open line sections Kőbánya-Kispest–Vecsés and the Cegléd–Szolnok "A" junction (both with 60E1 rail system). The difference is particularly striking between the latter open line section and the Abony through the track (60 rail), built simultaneously, where the gauge averages show a difference of 2 mm. It may be explained by the fact that the L2 sleepers were built with different dimensions so that a +2 mm gauge widening is included, while the "*LW-type*" sleepers were built with a 1435 mm gauge.

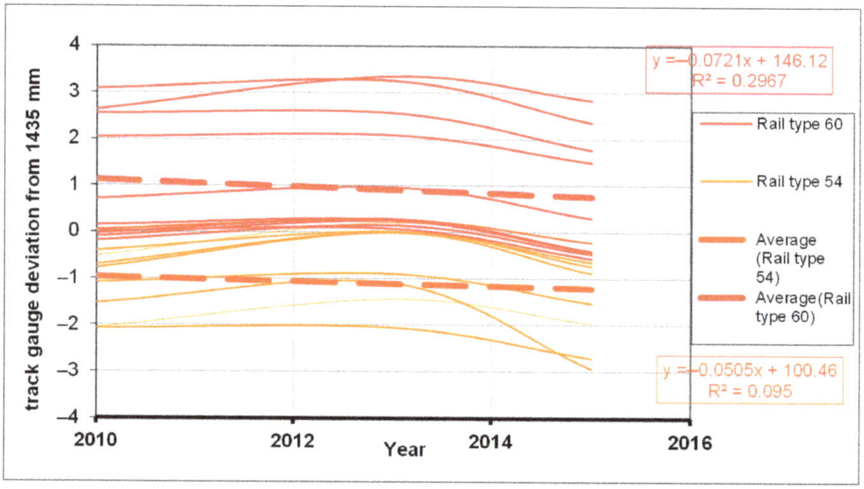

Figure 4. Change of gauge in straight sections of line 100a in the function of time (classified according to rail system).

2.7. Consequences of the Tests

Based on the authors' analysis, it can be concluded that gauge narrowing is typical for straight sections; it is not generated in curved tracks. It is because the lateral force occurring in curved sections generally causes rail wear, compensating for a possible narrowing of the track and gaps due to the dimensional tolerances of the rail fastenings, which act in the direction of the gauge widening.

The average growth rate of the gauge widening on curved sections for all rail sections on the studied lines is $\approx$0.5 mm/year. On straight track sections, the gauge of the old 48 rail-system superstructures narrowed at 0.15–0.25 mm/year over time.

Regarding gauge narrowing, the MÁV48.5 rail-system superstructure is the most affected. The explanation is that these superstructures are old, 50–55-year-old structures are common on the main lines, and the aging processes are already increasing. In straight sections, the gauge narrows as time progresses. The most affected are the concrete sleepers originally constructed with wooden inserts marked "*T-type*" and "*L-type*" and for the MÁV48 rail system, where the wooden inserts' aging and the water ingress into the dowels initiate a serious deterioration process.

The straight track sections with the UIC54 and UIC60/60E1 rail system are younger and did not or only slightly increase gauge narrowing over time.

3. Investigation of Gauge in Operational Tracks and in Laboratories

The authors conducted a field survey of the railway tracks regarding track gauge and rail profile at 15 track locations. From the on-site measurements of the track gauge, rail web, railbase, and ribs of baseplates, it was concluded that the most significant contributor to the development of gauge narrowing is the improper positioning (loss of distance) of the baseplate. It may be due to corroded wooden inserts in the cross sleepers or damage to the plastic dowels, possibly due to the expansion of the holes in the baseplate. This observation was particularly true for the 48 kg/m rail-system superstructure ("*L type*" cross sleepers).

Another but less significant cause of gauge narrowing is a change in the inclination of the rail. It was mainly observed in measuring the 54 and 60 kg/m rail-system superstructures but was insignificant in the 48 system. The change in rail inclination causes a maximum of 2 mm of gauge narrowing per rail by changing the rail inclination from 1:20 to approximately 1:16. More detailed measurements show that it is rare to see a marked increase in the cant of both rails.

The effect of increasing inclination of the rail was most evident in the 60 rail-system superstructure with direct flexible rail fastening, where the railpad under the rail was probably compressed asymmetrically, contributing to the development of gauge narrowing by 1-1 mm per each rail. However, it should be noted that such locations had the lowest effect on the total gauge narrowing of all the measured locations, which did not even reach the C1 warning limit.

The authors also carried out more measurements at three sites using instruments for the inclination of the rail and the rail profile itself. The site measurements confirmed the findings on rail inclination; the rail inclinations were not greater than 1:16, and a significant increase in inclination occurred in only one rail at a time.

Based on the field tests, 15 concrete sleepers were selected, removed from the track, and transported to the department's laboratory for more detailed testing. In the laboratory, the dimensions that influence the track gauge, such as the base lengths, distances between the ribs of the baseplates, and the inclination of the surface supporting the rail were measured.

At 100% of the *"L-type"* sleepers suitable for the MÁV48 rail system, originally with wooden inserts for the bolts, and at that time already 40–50 years old, it was observed that the baseplates could be moved by hand with an amplitude of about 5–7 mm due to the breaking of the double washers and the loosening of the bolts. This condition is unsuitable for ensuring gauge, and any gauge can develop. Theoretically, the movement could be as small as 1 mm (bore-to-screw diameter difference), but more significant movements may occur due to thinning of the screws due to corrosion and slight expansion of the holes. It is important to note that the rail bolt was adequately fixed in many cases, but the baseplate still moved on the sleeper due to the lack of clamping force. It is a conclusion that even with an intact dowel, the clamping effect is not always adequate, and this effect is difficult to detect in an assembled track.

4. Finite Element Modeling of Concrete Sleepers

4.1. Structure of the Model

A finite element (FE) model of concrete sleepers has been assembled to investigate how the effects during the manufacture of concrete sleepers, their installation in the track, and during operation affect the development of gauge narrowing.

For various finite element software packages, the modeling possibilities of the design and check calculation of reinforced concrete sleepers were investigated. The triaxial (three-axis) behavior of concrete can only be formulated using a continuum model. Using the ANSYS program, a basic model describing the behavior of concrete sleepers was developed, which describes the spatial deformation and stress state of prestressed reinforced concrete sleepers more accurately than previous models.

The geometric and strength data of the reinforced concrete sleeper were calculated based on the *"L4 SV-type"* sleeper used on the Hungarian railway network (see Figure 5). The length of the sleeper is 2.60 m, its width is 300 mm on the bottom surface, the height under the rail is 215 mm, the weight is 326 kg, and the class of strength of the concrete is C50/60.

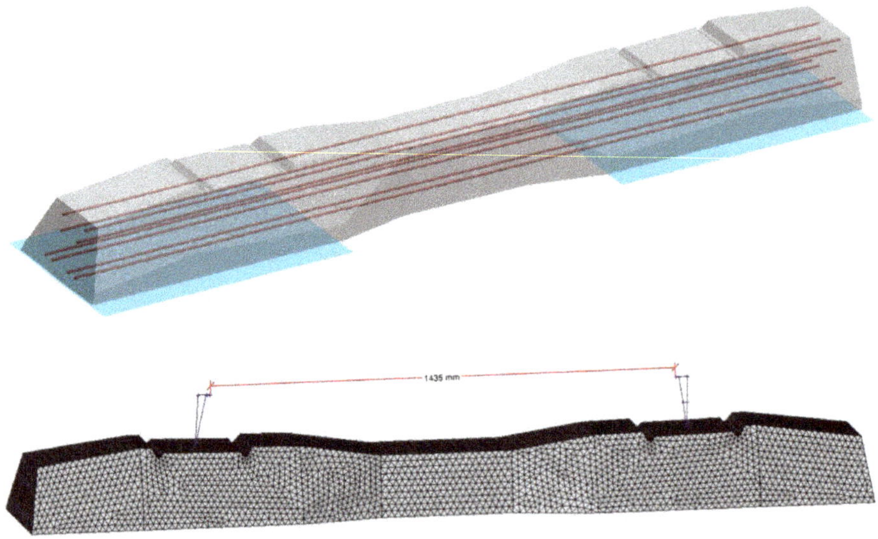

Figure 5. Schematic figures on the modeled *L4 SV-type* pre-tensioned reinforced concrete railway sleeper (above part: a 3D model with the steel tendons; below part: the applied mesh in the FE model).

The applied rail system is 60E1, its moment of inertia is $I = 30.55 \times 10^6$ mm^4. The inclination of the rail is 1:40.

Other relevant details of the FE model:

- The environmental categories of the concrete are XC4, XF3, and XA1 [37];
- There are eight pieces of steel tendons, type Y1770C-6.0-I, diameter 6.0 mm in accordance with [38];
- 69,000 body elements of the sleeper;
- 1000 rod elements (tendons);
- The sleeper was modeled by connecting the rail crown (top of the rail) finite element points in a rigid body (the height of the 60E1 rail profile is 172 mm; however, the rail gauge is measured 14 below the top of the rail and the rail inclination is 1:40, i.e., the height of the rigid body is $172 - 14 = 158$ mm as well as the width of the rail head at this height is 72 mm, i.e., the lateral dimension is $72/2 = 36$ mm; all these values are measured from the mid-point of the rail foot);
- The bolt holes for rail fastener screws were neglected;
- The initial tension of the steel tendons was 305 kN at the eight pieces; initial tensional stress is approximately 1350 MPa;
- The final stress value was considered with 1230 MPa;
- The final value of shrinkage 5.2×10^{-4} m, i.e., 0.52 mm;
- The creep factor was taken into consideration with 1.85.

4.2. Loads and Supports

For the vertical loads, an axle load of $Q = 250$ kN was assumed (it is the maximum axle load based on LM71 type vehicle load), equivalent to a wheel load of $Z = 125$ kN, and the speed was chosen to be $V = 160$ km/h (it is the maximum allowed speed in Hungary). Based on the principle of a longitudinal beam with continuous elastic bedding, the Zimmermann method was used with the Eisenmann dynamic factor to calculate the vertical load per sleeper. The sleeper spacing is $k = 600$ mm, and the surface of the half-sleeper on the ballast bed is $A = 850 \times 300$ mm^2, from which the width of the replacement longitudinal beam is $s = A/k = 425$ mm. The bedding factor can be $C = 0.05$ N/mm^3 for poor subsoil, $C = 0.10$ N/mm^3 for good subsoil, and $C = 0.11 - 0.15$ N/mm^3 for very good subsoil (all values are from [39–41]). The harder subgrade is the standard for the force transmitted from the railbase to the sleeper, so the authors' calculations used a value of C = 0.15 N/mm^3.

According to Eisenmann, the dynamic multiplication factor (β) can be calculated as follows [39–41]:

$$\beta = 1 + t \cdot \alpha \cdot \varphi \tag{3}$$

where for a medium quality superstructure δ = 0.2; for probability 0.997 t = 3 is assumed and φ is the speed-dependent factor (t must be 3.0 in the case of dimensioning) [39–41]:

$$\varphi = 1.0 + \frac{V-60}{140} = 1.0 + \frac{160-60}{140} = 1.714 \tag{4}$$

The value of the dynamic factor [39–41]:

$$\beta = 1 + t \cdot \delta \cdot \varphi = 1.0 + 3 \cdot 0.2 \cdot 1.714 = 2.03 \tag{5}$$

The force F transmitted to the sleeper from the rail according to the Zimmermann–Eisenmann function [39–41]:

$$F = \frac{k \cdot Z}{2 \cdot L_{lrb}} \cdot \beta \cdot \sum \eta \tag{6}$$

where L_{lrb} is the length of the longitudinal replacement beam [39–41]:

$$L_{lrb} = \sqrt[4]{\frac{4 \cdot E \cdot I}{C \cdot s}} = \sqrt[4]{\frac{4 \cdot 2.15 \cdot 10^5 \cdot 30.55 \cdot 10^6}{0.15 \cdot 425}} = 801.2 \text{ mm} \tag{7}$$

The value of the influence line of deflection is $\eta = e^{-\xi}(\sin\xi + \cos\xi) = 1$, where $\xi = x_{dist}/L_{lrb}$, x_{dist} is the absolute distance from the cross-section.

The load transmitted to the sleeper from the railbase [39–41]:

$$F = \frac{k \cdot Z}{2 \cdot L_{lrb}} \cdot \beta \cdot \sum \eta = \frac{600 \cdot 125}{2 \cdot 801.2} \cdot 2.03 \cdot 1 = 95.0 \text{ kN} \tag{8}$$

It should be noted that combining a medium-rated superstructure and a speed of 160 km/h is an extreme case; however, a very good subgrade was chosen as a boundary condition to maximize the force transmitted at the railbase. For this reason, the authors did not consider it appropriate to set the influence line of the deflection line to a value greater than 1.0 since this value will only be higher for locomotive bogies and only in exceptional cases for loaded trainsets.

The maximum value of horizontal lateral load (Y_{max}) based on previous experimental measurements [39–41]:

$$Y_{max} = 0.54 \cdot \left(10 + \frac{2}{3} \cdot Z\right) = 0.54 \cdot \left(10 + \frac{2}{3} \cdot 125\right) = 50.4 \text{ kN} \tag{9}$$

Of this load, approximately 65–70% is conveyed onto one sleeper ($Y_{max}^{sleeper}$)—it is because the results of load distribution based on the Zimmermann-calculation method in the worst case (see [39–41]):

$$Y_{max}^{sleeper} = 0.65 \ldots 0.70 \cdot Y_{max} = 33 \ldots 35 \text{ kN} \tag{10}$$

Based on the above, the forces from the wheel loads of the vehicles acting on a rail support surface of the sleeper are calculated to be 95 kN vertically and 35 kN horizontally with the point of application of the horizontal force 14 mm below the railtop level.

In the modeling of concrete sleepers, two types of load transfer were investigated: one is centric, and the other is eccentric. In the case of concentrated loading, the sleepers and the ballast are adequately supported and maintained, with a half-sleeper bearing surface of

$A = 850 \times 300$ mm^2 on the ballast bed (Figure 6). In Figures 6 and 7, the red pyramid is an infinite rigid mesh, which means the uniform distribution of the concentrated force on the surface on the bearing surface under the rail in the FEM model.

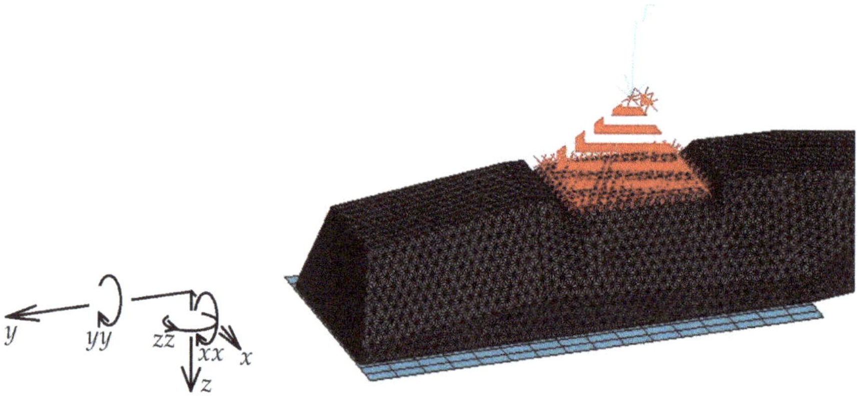

Figure 6. Central loads and supports in normal operation (the red–white tetrahedron symbolizes the load distribution from the rail head to rail foot to be able to consider realistic loading onto the modeled sleeper). Below the sleeper, a surface was considered to support the sleeper; see the meshed "plate".

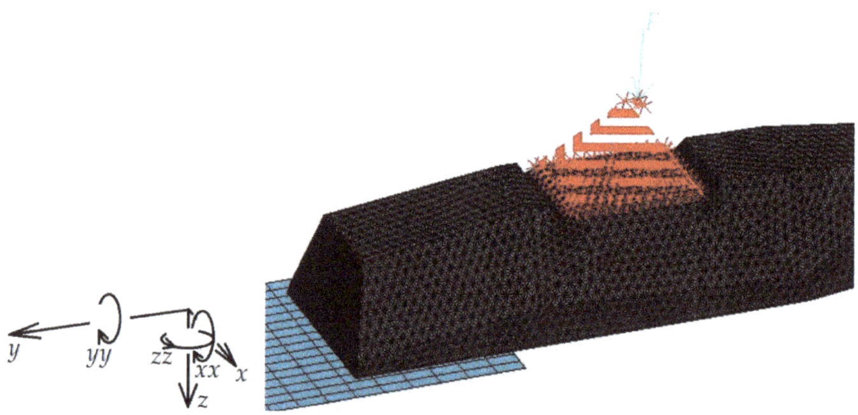

Figure 7. Eccentric loads and supports (the red–white tetrahedron symbolizes the load distribution from the rail head to rail foot to be able to consider realistic loading onto the modeled sleeper). Below the sleeper, a surface was considered to support the sleeper; see the meshed "plate".

However, the effect of uneven support of the sleepers (uneven tamping, tamping only the part of the sleeper outside the rail, water bag, etc.) on the deflection of the track was also investigated. The case was modeled where the wheel load is not transmitted uniformly from the sleeper to the ballast but eccentrically, and the sleeper is supported on the ballast for only 300 mm long from the ends of the sleepers due to uneven tamping or highly inhomogeneous loadbearing of the substructure (Figure 6). In the case of eccentric loading, the point of application of the vertical load is in the third of the railbase towards the centreline of the track, and the sleepers are supported only over a length of 300 mm from the end of the sleeper, with a support area of 300×300 mm. The central and eccentric support models are illustrated in Figures 6 and 7.

In the modeling, the effects of concrete shrinkage, creep, and relaxation of the prestressing wires were also taken into account.

4.3. Cases of Assessment

The tests should be analyzed from the perspective of the amount of gauge narrowing caused by changes in displacement boundary condition (support) and by loads, taking into account rheological changes. In all cases, the initial gauge was 1435 mm, and the modeling was carried out relative to this. The following cases were investigated:

(a) 50 days; relaxation, creep, shrinkage;
(b) 100 days; relaxation, creep, shrinkage;
(c) 400 days; relaxation, creep, shrinkage;
(d) Same as (c) + operational central loads;
(e) Same as (c) + operational eccentric loads;
(f) Same as (e) + 35 °C cooling of temperature;
(g) Same as (e) + 10% over-tensioning of the wires;
(h) 400 days; eccentric loads, self-weight.

There are further details for the above cases. The tests should be viewed from the perspective of the amount of compression caused by changes in displacement boundary condition (support) and force, taking into account rheological changes. The age of the concrete sleeper is also given when considering the rheological effects.

In these investigations, it was assumed that the rails are placed on the concrete sleepers at 50 or 100 days of age, at which time the railway track geometry is adjusted; i.e., it will take around 50–100 days from production to installation. During this time, the sleepers will have a length change caused by tension, relaxation, creep, and shrinkage. These are the baseline data, i.e., the so-called "zero" values regarding pressure shrinkage. Here, they are controlled (adjusted), and the railway can run without any rail gauge narrowing. Over time, however, the sleepers suffer further length changes, which are corrected only by repeatedly correcting the pressure contraction determined by the mandatory periodic inspections. This date was taken in the current study to be the 400th day after the manufacture of the sleepers.

Based on the above-mentioned facts, the shrinkage of the concrete was taken into account in all cases (see Figures 8 and 9) when determining our calculation cases, and the same accounting was performed for relaxation and creep.

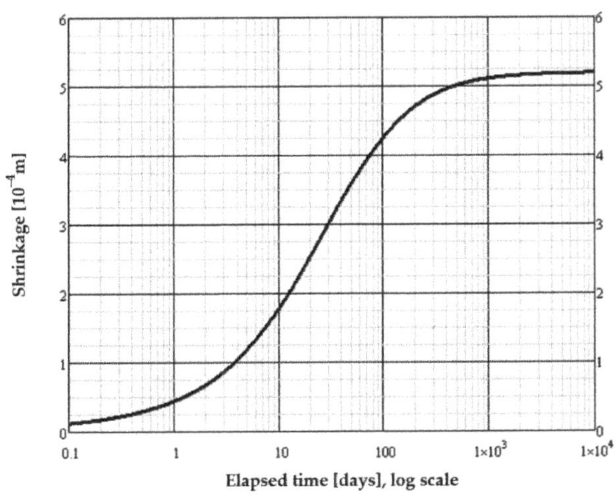

Figure 8. Calculated shrinkage of the concrete until the first 10,000 days after manufacture.

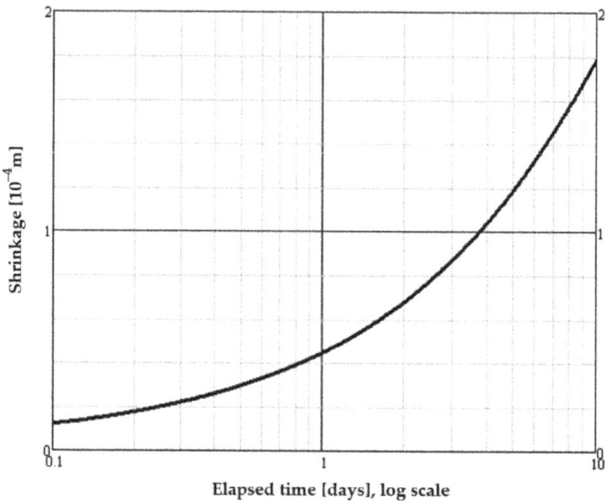

Figure 9. Calculated shrinkage of the concrete until the first ten days after manufacture.

The method used to determine the creep rate is based on Section 3.1.4 of EN 1992-1-1 [42]. Assuming an outdoor environment, 0 days of prestressing, and the designed concrete strength class, the creep factor was found to be 1.85. Consequently, the long-term modulus of elasticity of the concrete will be equivalent to a value of (1 + 1.85). The actual concrete structure (real concrete mix) has a much higher strength. It only reduces the creep, so our estimate will be upper bound.

The relaxation of the prestressing wire was calculated according to Section 3.3.2 of the literature [42], assuming a stress loss of 2.5% at 1000 h. The manufacturer provides this latter value, but it is the guaranteed value for modern reduced relaxation wires (tendons) used today. Note, however, that this value could be as high as 8% for old conventional relaxation wires.

4.4. Results

Figure 10 illustrates the results of case (a). Due to the limited space, the authors decided not to publish all the details with similar figures related to FE modeling.

Table 3 summarizes the results of the computer runs (parametric tests). The cases represent the combination of effects considered. The gauges and the value of the narrowing are given separately.

Table 3. Values of gauge narrowing.

Load Case (See Section 4.3)	Description of Load Case	Gauge [mm]	Narrowing of Gauge [mm]
(a)	50 days; relaxation, creep, shrinkage	1434.39	0.61
(b)	100 days; relaxation, creep, shrinkage	1434.24	0.76
(c)	400 days; relaxation, creep, shrinkage	1434.00	1.00
(d)	Same as *c*; +operational central loads	1433.95	1.05
(e)	Same as *c*; +operational eccentric loads	1427.00	8.00
(f)	Same as *e*; +35 °C cooling of temperature	1426.62	8.38
(g)	Same as *e*; +10% over tensioning of the wires	1426.46	8.54
(h)	400 days; eccentric loads, self-weight	1433.78	1.22

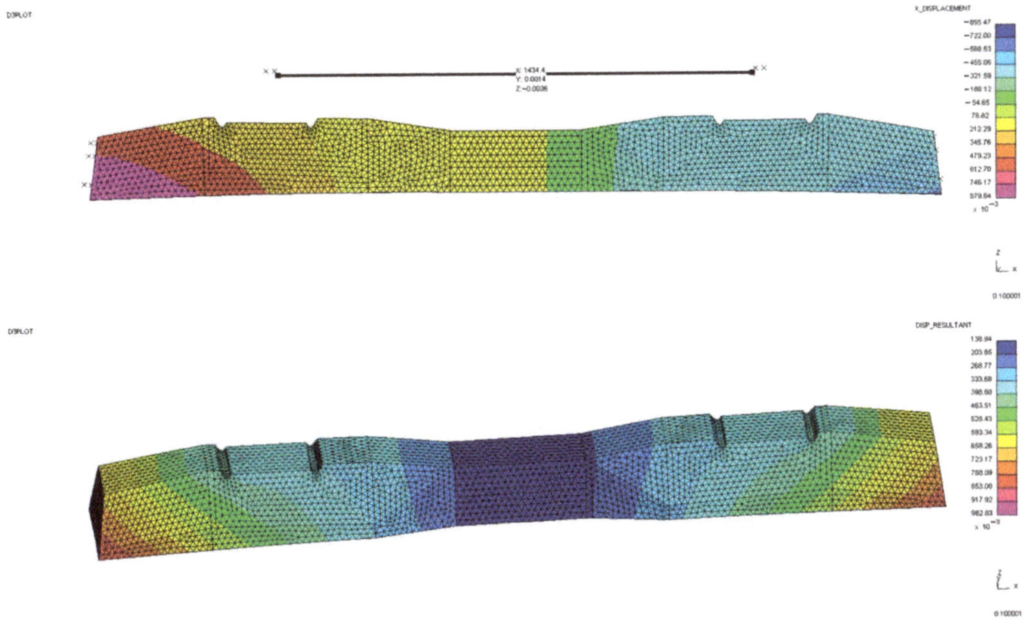

Figure 10. The *x-direction* (**above part**) and total displacement (**below part**) plots, in the case of (a), are based on Section 4.3. The legend shows values with 10^{-3} mm units.

The calculations were based on the assumption that after the sleepers are manufactured, they are stored at the manufacturer's site, then transported, possibly also stored before installation, then installed, the rails are placed, and their track gauge is adjusted. From production to the track alignment takes about 50–100 days. During this time, the sleepers will have a length variation caused by tension, relaxation, creep, and shrinkage. These values are shown in rows '(a)' and '(b)' of Table 3. These are the initial baseline data, i.e., the "0" (zero) values for the gauge narrowing. It is because the track and gauge are controlled here, and the railway can run without narrowing. Over time, however, the sleeper suffers further length variations, determined by the mandatory periodic inspections and corrected only by repeated maintenance works. Therefore, this date was taken in the authors' tests to be the 400th day after manufacture.

At 50 days of age, the gauge narrowing (row "a" in Table 3) is 0.61 mm. This value can also be taken as a baseline, the "0" (zero) value. At 100 days of age, the gauge narrowing (row 'b') is 0.76 mm, which can, of course, also be taken as a baseline if someone assumes that the rail positioning and gauge is set on this day; i.e., this value can also be taken as '0' for the gauge narrowing. The track gauge can be calculated from case 'a' or 'b' for the first time because this is the first time the track gauge is adjusted during construction works, and there is no narrowing at this time; the track gauge is still 1435 mm.

At 400 days of age, the gauge narrowing from production (Table 3, line "c") is 1.00 mm. Therefore, the gauge narrowing determined by the mandatory periodic supervision tests is assumed to be corrected by realigning at about 400 days from manufacture. In other words, if someone is looking for the value of the gauge narrowing between the 50th and the 400th day, a value of

$$N(a) = c - a = 1.00 - 0.61 = 0.39 \text{ mm} \tag{11}$$

is received, and if someone looks for the value of the gauge narrowing between day 100 and day 400, then a value of

$$N(b) = c - b = 1.00 - 0.76 = 0.24 \text{ mm} \tag{12}$$

is received.

These values are, of course, derived only from the sleepers' deformations without the trains' loads.

The load combination 'd' in the table shows the displacement results calculated from numerical modeling of the combination of relaxation, creep, shrinkage, and the in-service load for 400 days, with central wheel loads and central support on the ballast. In this case, the gauge narrowing of the sleeper was 1.05 mm from the date of manufacture. If the narrowing is calculated from the age of 50th or 100th days, the former values are as follows:

$$N(d - 50) = d - a = 1.05 - 0.61 = 0.44 \text{ mm} \tag{13}$$

or

$$N(d - 100) = d - b = 1.05 - 0.76 = 0.29 \text{ mm} \tag{14}$$

Significantly higher values are obtained if eccentric wheel loads are considered, i.e., if the support of the sleeper is only about 30–30 cm long at both ends and the vertical loads are also eccentric. In this case, the rate of gauge narrowing is 8.00 mm from the 0 date of manufacture, and considering it from the 50th or 100th day after manufacture, it is

$$N(e - 50) = e - a = 8.00 - 0.61 = 7.39 \text{ mm} \tag{15}$$

or

$$N(e - 100) = e - b = 8.00 - 0.76 = 7.24 \text{ mm} \tag{16}$$

These values are very significant compared to the normal case "d". They do not occur in the case of normal maintenance of the track and ballast but only in the case of extreme tamping and ballast deficiencies. It clearly highlights the need for professional track maintenance.

If, in addition to the effects of relaxation, creep, shrinkage, and the combination of external loads calculated for 400 days, a 35 °C cooling is included, the former values increase by a further 0.38 mm.

For an overstressing of the tension wires by 10%, the additional increase in gauge narrowing is 0.54 mm compared to case "e".

4.5. Conclusions

Finite element modeling has shown that the sleeper's shrinkage and other time-dependent deformation do not contribute significantly to the development of the gauge narrowing. The maximum magnitude of the time-dependent effects remains below 1.0 mm. With time-dependent effects, central transfer of wheel loads, and proper support of the cross sleepers—careful maintenance—the cumulative rate of gauge narrowing is also around or below 1 mm, i.e., insignificant.

However, the uneven support of the cross sleepers and the eccentric transfer of wheel loads to the sleeper can cause a gauge narrowing of up to 7.39 mm. The cumulative values of these effects can lead to significant gauge narrowing of 8–9 mm. With careful design and maintenance, high narrowing values can be avoided by ensuring adequate and uniform support of the concrete sleepers. The results are summarized in Figure 11.

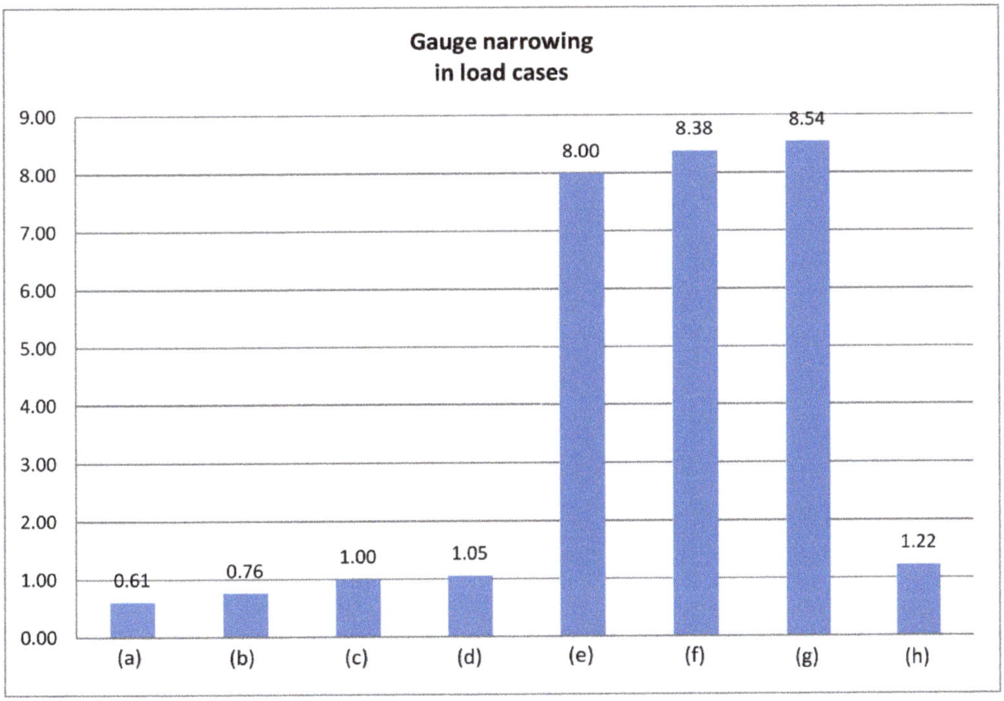

Figure 11. Result of gauge narrowing (all units are in mm) in the different load conditions (the legend from (a) to (h) are in accordance with Section 4.3).

5. Modeling of Lateral Stability of Rails under Loss of Function of Rail Fastenings

In this R&D work, the authors are trying to find answers to the question of what factors influence and what causes lead to the formation of a gauge narrowing. Taking into account the rail–wheel interaction, the following forces act on the rail:

- Vertical forces resulting from the static and dynamic wheel loads of the vehicles;
- Horizontal lateral forces from the irregular movements of the vehicles and non-compensated lateral accelerations;
- Longitudinal forces from breaking and accelerations of the vehicles and the temperature change in case of constrained heat expansion.

The track gauge is most affected by the lateral force, but it acts outward from the track axis, which also explains the phenomenon of wear on the leading edge of the rail. The present study shows that in curves, there is never a contraction of the gauge, but rather a widening of the gauge, which, depending on the radius of the curve, increases at a rate of 0.5–1 mm/year. However, a small or large degree of gauge narrowing occurs on straight track sections, especially on old, aged track sections.

Rail fastenings limit the movement of the rail normally in all six degrees of freedom on the sleeper (see Figures 6 and 7):

- Longitudinal rail restraint in the *x-direction;*
- Lateral support of rail in the *y-direction* and keeping the gauge;
- Vertical support of the rail in the *z-direction*, that is, to bear wheel loads;
- Restrict rotation in the *xx-direction* by two rail clips equally holding down the rail;
- Restrict rotation in the *yy-direction;*
- Restrict rotation in the *zz-direction*, that is, the torsional stiffness of the fastening to serve for the stability of the track against buckling.

Of the effects listed above, the variation in gauge is primarily caused by insufficient support in the *y-direction*. However, twisting in the *xx-direction* can also cause a gauge error since the railbase is supported, and the gauge is measured at the rail head. A change in the height of the railbase on either side also causes a displacement of the rail head.

The rotation in the *zz-direction* will cause a *y-directional* shift in the adjacent rail supports.

5.1. Thermal Forces in Case of Constrained Heat Expansion

In a continuously welded rail track (CWR track), the following normal force is generated if the temperature of the track is not equal to the actual neutral temperature [39–41]:

$$N = \alpha \cdot E \cdot A \cdot \Delta T \tag{17}$$

For example, the compressive force developed in the 54E1 and 60E1 rails for temperature increases of ΔT = 45 and 50 °C, respectively, are given in Table 4.

Table 4. Compression force in the 54E1 and 60E1 rails due to an increase of temperature of 45 °C and 50 °C.

Increase of Temperature ΔT [°C]	Compression Force F [kN]	
	54E1 Rail	**60E1 Rail**
45	776.28	853.38
50	862.53	948.20

In the case of standard track design, the actual neutral temperature at the time of construction should be set between +15 and +28 °C (neutral temperature zone) according to the Instruction of D.12/H "Instruction for the Construction, Maintenance and Supervision of CWR Tracks" of MÁV Zrt. [43]. Of course, the neutral temperature in the track can vary from place to place and is constantly affected by the displacement of the sleepers and longitudinal forces (e.g., by acceleration or braking of the trains or inhomogeneous ballast conditions, etc.). Maintenance works in the track (repair of rail fracture, replacement of a section of rail, turnout replacement, etc.) can all change the original neutral temperature. At certain time intervals, it may be necessary to release the tensile stresses in the rail and establish a new effective neutral temperature on a more extended section.

5.2. The Euler Force

From the point of view of buckling, the compressive force at high temperatures is of interest. The minimum critical force F_{cr} at which the rail loses its stability without lateral support should be investigated for the above reasons. According to Euler's theory of stability, the critical force F_{cr} is the force at which the rail is in a state of equilibrium even in the deflected shape [39,40]:

$$F_{cr} = \frac{\pi^2 \cdot E \cdot I_y}{\lambda^2 \cdot l_{ws}^2} \tag{18}$$

In the present case, both ends of the rail are assumed to be fully fixed, i.e., λ = 0.5. The computational results are summarized in Table 5 under the following assumptions:

- 1…16 rail fastenings do not have sufficient lateral support for the rail; beyond them, all fastenings are satisfactory;
- The cross-section of the rail cannot take a force greater than the force calculated from the yield strength and the cross-section of the rail, so the values in the table include not only the values calculated from the above formula but also the values calculated from the yield strength σ_f = 500 MPa. These values are highlighted in red.

- The critical force F_{cr} is greater than the thermal force in the rail in the range of smaller deflection lengths, so the rail remains stable under the maximum thermal compression force for deflection even if nine fastenings do not work laterally for 54E1 rails and 10 for 60E1 rails. The force F_{cr} calculated, in this way, cannot develop in the rail due to temperature changes. These values are highlighted in yellow in the table.

Table 5. The critical force F_{cr} belonging to different rail sections and unsupported lengths of rail.

Number of Rail Fastenings That Have Failed n [—]	Length of Rail without Lateral Support l_{ws} [mm]	Critical Force F_{cr} [kN]	
		54E1	60E1
1	1200	3488.50	3835.00
2	1800	3488.50	3835.00
3	2400	3488.50	3835.00
4	3000	3488.50	3835.00
5	3600	2745.46	3355.19
6	4200	2017.07	2465.04
7	4800	1544.32	1887.30
8	5400	1220.20	1491.20
9	6000	988.36	1207.87
10	6600	816.83	998.24
11	7200	686.36	838.80
12	7800	584.83	714.72
13	8400	504.27	616.26
14	9000	439.27	536.83
15	9600	386.08	471.82
16	10,200	341.99	417.95

If 10 fastenings fail for 54E1 rails and 11 fastenings fail for 60E1 rails laterally, the force generated in the rail by the temperature rise exceeds the critical compressive force for buckling, and the rail is not stable.

5.3. Investigation of Rail Length Variation as a Function of Initial Misalignment

At temperatures higher than the actual neutral temperature, the compressive force in the rail is generated due to the inhibited length change. It also follows that if the length change is allowed, the compressive stress will decrease. Therefore, it is also necessary to consider whether the length change of the rail due to initial misalignment represents an amount of length change associated with a degradation of the stress. From the calculations, it is clear that in the case of a type A failure mode for different wavelengths and amplitudes, the change in length of the rail is negligible for amplitude values of initial misalignments. The change in length of rail due to the wavelength-amplitude pairs that can be formed is only 0.01–0.02 mm, which does not affect the thermal force and is only a few thousandths of the compression due to the thermal force.

5.4. Investigation of Rail Length Variation as a Function of Initial Misalignment

It can be seen that in some cases—in case of inadequate lateral support of 9–11 fastenings—the rail may suffer lateral buckling. Therefore, examining the relationship between the longitudinal compressive force and the buckling is necessary. The theoretical straight rail is not applicable for such an analysis because the laws of mechanics state that the rail is in equilibrium, and only the critical force F_{cr}, which is associated with the loss of stability, as stated earlier, can be determined. The deflected shape cannot be determined.

The test is therefore carried out by assuming a distorted geometry of the rail, which can be modeled by a slight initial misalignment with an amplitude of f and with a length of l_{dist} and a type A. The shape of the wave is a half-cosine wave. Based on this, the function of initial geometrical distortion, in the function of x, is [39,40]

$$y_k(x) = \frac{y_{k,max}}{2}\left(1 - \cos\left(\frac{2\cdot\pi}{l_{dist}}x\right)\right) \tag{19}$$

The longitudinal force causes a bending moment in the structure, given that the shape of the beam is not straight. The magnitude of the bending moment as a function of x [39,40] is

$$M(x) = F_{comp}\cdot y_k(x) = F_{comp}\cdot y_{k,max}\left(\cos\frac{2\cdot\pi}{l_{dist}}x\right) \tag{20}$$

According to the differential equation of the elastic beam, the applied bending moment is proportional to the curvature of the support, which is related to the following [39,40]:

$$\frac{1}{\rho(x)} = \frac{M(x)}{E\cdot I_y} \tag{21}$$

Based on the differential equation of the elastic beam, the curvature of the support can be used to determine the deflected shape of the beam as follows [39,40]:

$$\tau(x) = \int_0^x \frac{M(x)}{E\cdot I_y}dx \tag{22}$$

$$y(x) = \int_0^x \tau(x)dx \tag{23}$$

After the integrations and substituting $x = l/2$ [39,40],

$$y(x) = \frac{F\cdot y_{k,max}}{2\cdot E\cdot I_y}\cdot\frac{{l_{dist}}^2}{4\cdot\pi^2}\left(1 - \cos\left(\frac{\pi}{l_{dist}}x\right)\right) \tag{24}$$

The maximum of $y(x)$ occurs when the bracketed value is maximum. It occurs when $\cos\left(\frac{\pi}{l}x\right) = -1$, i.e., $x = l/2$. The function, therefore, reaches its maximum at half of the error length considered [39,40]:

$$y_{max} = \frac{F\cdot y_{k,max}}{E\cdot I_y}\frac{l_{dist}^2}{4\cdot\pi^2} \tag{25}$$

By substituting actual values into the formula, it can be seen that y_{max} is greater than $y_{k,max}$, so the beam will bend due to the bending moment from the initial distortion but will consequently experience an even more significant bending moment, causing it to bend further. The equilibrium position is reached when the bending moment from the deflected shape is in equilibrium with the geometry of the deflected shape, i.e., [39,40]

$$y_{max} = \frac{F\cdot(y_{k,max} + y_{max})}{E\cdot I_y}\frac{{l_{dist}}^2}{4\cdot\pi^2} \tag{26}$$

In the above equation, it is assumed that the initial deflection $y_{k,max}$ does not include the bending moment in the rail; otherwise, the problem can be reduced to the case of a straight, free-of-distortion girder, and the actual deflection y_{max} associated with the longitudinal force cannot be determined.

The y_{max} deflection can also be determined by iteration. The maximum deflection y_{max}, derived from the bending moment *M(x)* associated with the initial deflection $y_{k,max}$, is substituted back into the initial deflection value by increasing it by the original initial deflection $y_{k,max}$.

5.5. Analysis of the Force Greater Than the Critical Force in Case of Partial Lateral Support

The rail is held down by the rail fastenings in the *x*-, *y*- and *z-directions*; the standard sleeper spacing on main lines is 0.60 m. Based on the authors' field and laboratory tests, the experience is that the baseplates of the rail fastenings move laterally so that the lateral support of the rail is incomplete within certain limits. After an initial *y-directional* displacement, the baseplates are supported, but until then, their movement is possible within certain limits.

As shown in the previous chapters, if the compressive force exceeds the critical force F_{cr}, the track loses its stability without lateral support. However, as the baseplates, which are not provided sufficient lateral support, can only move freely up to a certain extent and then hit the rail bolt, lateral support can be provided but with a narrower track gauge. Thus, another equilibrium condition can be created, which, assuming a straight track, cannot be changed by the *z-direction* load of the railway vehicle.

At this point, the deflected shape will no longer be a pure sine curve because the lateral support of the rail may disturb it.

The support in the 9 m long model (with 16 fastenings) is assumed to be at the two middle sleepers (k = 0.60 m sleeper spacing), each with a force F_H (to achieve symmetry), as shown in Figure 12. The bending moment due to the supporting forces on the support is indicated by the green line in Figure 13. There are also bending moments at both fixed ends of the rail, the magnitude of which can be calculated from the boundary conditions.

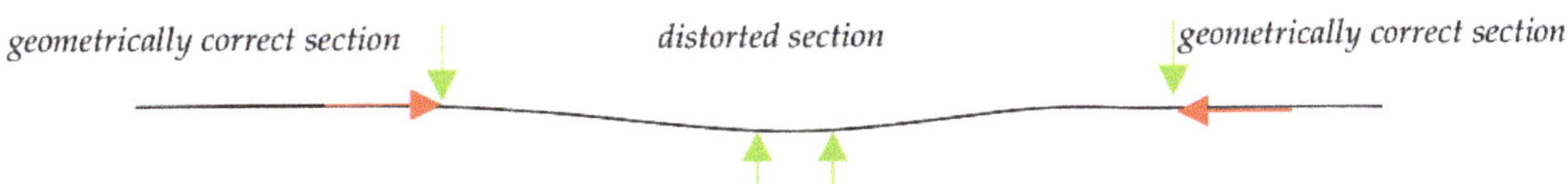

Figure 12. Forces acting on the rail with initial distortions (l_{dist} = 9 m, 54E1; the horizontal arrows symbolize the longitudinal forces in the track; hence, the vertical arrows mean the transversal forces).

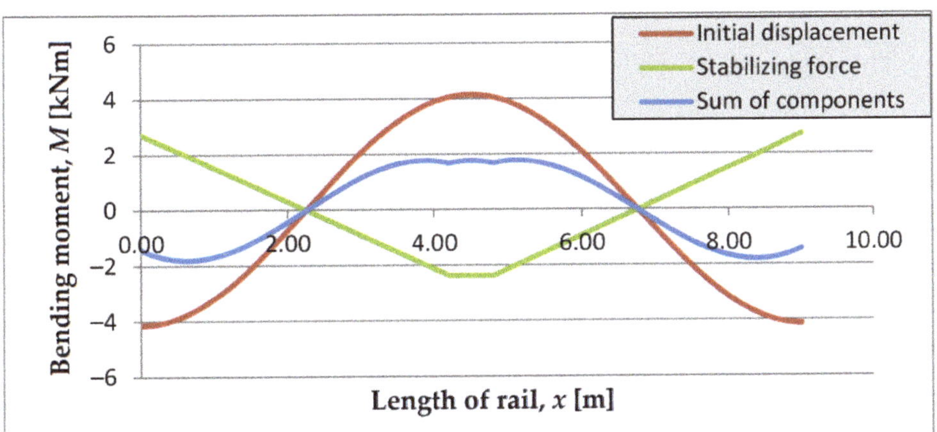

Figure 13. Bending moment in the rail with initial distortions (l = 9 m, 54E1).

The differential equation of the elastic beam is [39,40]:

$$\frac{1}{\rho(x)} = \frac{M(x)}{E \cdot I_y} \tag{27}$$

The angle of the tangent of the rail at the length of x [39,40]:

$$\tau(x) = \int_0^x \frac{M(x)}{E \cdot I_y} dx \tag{28}$$

Since both ends of the rail are fully fixed, they cannot rotate; therefore, $\tau(0) = 0$ and $\tau(l) = 0$ [39,40]:

$$\tau(l) = \int_0^l \frac{M(x)}{E \cdot I_y} dx = \frac{(l_{ws} + k) \cdot (l_{ws} - k) \cdot F_H}{2 \cdot E \cdot I_y} - \frac{M_F \cdot l_{ws}}{E \cdot I_y} = 0 \tag{29}$$

Consequently, the value of the bending moment at the two fixed ends of the rail [39,40] is

$$M = \frac{(l_{ws} + k) \cdot (l_{ws} - k) \cdot F_H}{4 \cdot l_{ws}} = 2.24 \cdot F_H \tag{30}$$

The authors are investigating the value of the force F_H that can stabilize the track to a specified deflection value for a given initial deflection and a given longitudinal force. Since the track section is assumed to be symmetric, the lateral displacement will be the largest in the middle cross-section. Furthermore, an iterative numerical solution to the problem has been obtained since the function of the bending moment from the stabilizing force F_H consists of several sections.

First, the function is determined by numerical integration, and from this—also by numerical integration—the lateral displacement function $y(x)$. An example of the displacement function is shown in Figure 14.

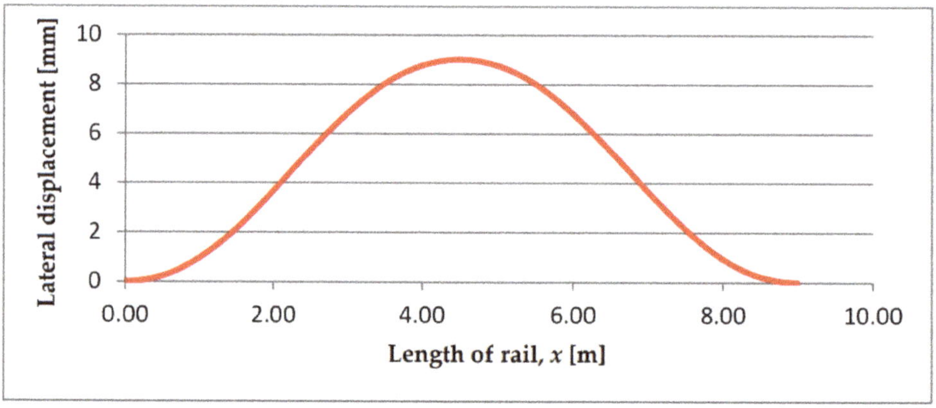

Figure 14. Lateral displacement of the examined section of rail.

As seen in the previous chapters, the rail section is inherently stable for a force lower than the thermal force F_{cr}, so in this case, a compressive force (thermal force) higher than the force F_{cr} must be assumed. The structure becomes unstable for such a high compressive force, which is realized in increased lateral displacement beyond all limits.

If this is prevented by the applied lateral force F_H, the lateral displacement is stopped, and an equilibrium situation occurs. In the case of the present study, this occurs when the rail bolt or rib of the baseplate stops movement and can no longer shift. It will depend on the condition and dimensional tolerance of the anchors in the concrete sleeper.

In the present study, not only an initial error is considered, but also specific values of lateral displacement are taken into account (which can be formed in any case due to the high thermal force), and then an iteration method is used to find the magnitude of the force that can stabilize the system.

- The size of the initial imperfection was set to the minimum ($y_{k,max}$ = 1 mm) but was not neglected because of the consideration that there are always slight imperfections in a long, thin fabrication. No bending moment is generated in the rail from a support; its shape is bent by the initial defect, and the rail is stress-free.
- The lateral displacement was found to be a function of the wavelength of the distortion ($y_{max} = l/1000$). It is because large lateral displacements cannot occur for small wavelengths, simply just due to the inherent bending stiffness of the rail.
- The values of length of distortions ($l_{dist} \leq 3.0$ m), which were stable in the authors' previous tests, are out of the scope of investigation now; no lateral support is needed here, and the rail section is stable even under the effect of the maximum thermal force.

The results are shown in Tables 6 and 7. A positive value means the rail is stable, so it needs extra compression force (active force) to reach the required deflection value in the middle of the distortion. These values are, therefore, irrelevant because no such force is active in the track. However, for the cells colored red, the result is relevant, showing the lateral support required to prevent buckling.

Table 6. Lateral support requirements for rail 54E1 [kN] for different lengths of distortions and longitudinal compressive forces in the case $y_{k,max}$ = 1 mm, and $y_{max} = l/1000$.

ΔT [°C]	F [kN]	Wavelength of Distortion [m] 3.6	4.2	4.8	5.4	6.0	6.6	7.2	7.8	8.4	9.0
5	83.7	6.26	4.81	3.42	2.75	2.11	1.75	1.40	1.18	0.98	0.83
10	167.5	6.00	4.54	3.18	2.50	1.87	1.51	1.17	0.95	0.75	0.60
15	251.2	5.74	4.27	2.93	2.25	1.64	1.27	0.93	0.71	0.52	0.37
20	334.9	5.48	4.01	2.68	2.00	1.40	1.03	0.70	0.48	0.29	0.14
25	418.6	5.22	3.74	2.44	1.75	1.16	0.79	0.47	0.24	0.06	−0.09
30	502.4	4.96	3.47	2.19	1.50	0.92	0.55	0.24	0.01	−0.17	−0.32
35	586.1	4.70	3.20	1.95	1.25	0.69	0.31	0.01	−0.22	−0.40	−0.55
40	669.8	4.44	2.94	1.70	1.00	0.45	0.07	−0.22	−0.46	−0.63	−0.78
45	753.6	4.18	2.67	1.45	0.75	0.21	−0.17	−0.46	−0.69	−0.87	−1.01

Table 7. Lateral support requirements for rail 54E1 [kN] for different lengths of distortions and longitudinal compressive forces in the case $y_{k,max}$ = 2 mm, and $y_{max} = l/1000$.

ΔT [°C]	F [kN]	Wavelength of Distortion [m] 3.6	4.2	4.8	5.4	6.0	6.6	7.2	7.8	8.4	9.0
5	83.7	6.21	4.75	3.38	2.71	2.08	1.71	1.37	1.15	0.96	0.80
10	167.5	5.89	4.44	3.09	2.42	1.80	1.44	1.11	0.89	0.70	0.55
15	251.2	5.57	4.12	2.80	2.13	1.53	1.17	0.85	0.63	0.45	0.30
20	334.9	5.25	3.80	2.51	1.85	1.26	0.90	0.59	0.37	0.19	0.05
25	418.6	4.94	3.48	2.23	1.56	0.99	0.63	0.33	0.11	−0.06	−0.20
30	502.4	4.62	3.16	1.94	1.27	0.72	0.36	0.07	−0.15	−0.32	−0.46
35	586.1	4.30	2.85	1.65	0.98	0.45	0.09	−0.19	−0.41	−0.57	−0.71
40	669.8	3.99	2.53	1.36	0.69	0.18	−0.19	−0.45	−0.67	−0.83	−0.96
45	753.6	3.67	2.21	1.07	0.40	−0.10	−0.46	−0.71	−0.93	−1.09	−1.21

The calculated values are small so that the rail fastening can support the rail laterally, at least at the calculated distortion lengths. However, the "wobbly", manually movable baseplates cannot even provide this force, only in their end position, so the rail moves freely between the end positions of the baseplate due to the thermal force, causing a gauge narrowing.

6. Conclusions

The results and conclusions drawn thereof are summarized as follows.

Analysis of the results of the measurement trains shows that gauge narrowing is typical for straight sections; it is not generated in curved tracks.

- On straight track sections, the track gauge of the old 48 kg/m rail-system superstructures has narrowed by 0.15–0.25 mm/year over time. Regarding gauge narrowing, the MÁV48.5 rail-system superstructure is the most affected. The explanation is that these superstructures are old, and 50–55-year-old structures are quite common on the main lines, and the aging processes are already increasing. In straight sections, the gauge narrows as time progresses. The most affected are the concrete sleepers originally built with wooden inserts marked *"T-type"*, and *"L-type"* and for MÁV48 rail system, where the aging of the wooden inserts and the ingress of water into the dowels initiate a serious deterioration process.
- Finite element modeling has shown that the sleeper's shrinkage and other time-dependent deformation do not contribute significantly to the development of the gauge narrowing. The maximum magnitude of the time-dependent effects remains below 1.0 mm. With time-dependent effects, central transfer of wheel loads, and proper support of the cross sleepers—careful maintenance—the cumulative rate of gauge narrowing is also around or below 1 mm, i.e., insignificant. However, the uneven support of the cross sleepers and the eccentric transfer of wheel loads to the sleeper can cause a gauge narrowing of up to 7.39 mm. The cumulative values of these effects can lead to significant gauge narrowing of 8–9 mm. With careful design and maintenance, high narrowing values can be avoided by ensuring adequate and uniform support of the concrete sleepers.
- In the case of loss of function of the rail fastenings, if 10 fastenings fail laterally for 54E1 rails and 11 fastenings fail for 60E1 rails, the force generated in the rail by the temperature rise exceeds the critical compression force for buckling, and the rail is not stable. However, for shorter lengths without support, the rail remains stable. Baseplates that do not provide sufficient lateral support can move freely only to a certain extent and then hit the rail bolt, thus providing lateral support—but with a narrower gauge. With an initial misalignment of 1 mm and the wavelengths and compressive forces in the rail shown in Tables 6 and 7, it was found that even small lateral forces can stabilize the rail and support it laterally.

Author Contributions: Conceptualization, P.B., N.L., Á.V. and S.F.; methodology, P.B., N.L., Á.V. and S.F.; software, P.B., N.L., Á.V. and S.F.; validation, P.B., N.L., Á.V. and S.F.; formal analysis, P.B., N.L., Á.V. and S.F.; investigation, P.B., N.L., Á.V. and S.F.; resources, P.B., N.L., Á.V. and S.F.; data curation, P.B., N.L., Á.V. and S.F.; writing—original draft preparation, P.B., N.L., Á.V. and S.F.; writing—review and editing P.B., N.L., Á.V. and S.F.; visualization, P.B., N.L., Á.V. and S.F.; supervision, P.B., N.L., Á.V. and S.F.; project administration, P.B., N.L., Á.V. and S.F.; funding acquisition, P.B., N.L., Á.V. and S.F. All authors have read and agreed to the published version of the manuscript.

Funding: This research received no external funding.

Data Availability Statement: Not applicable.

Acknowledgments: The authors would like to thank MÁV Ltd. and MÁV CRTI Ltd. for their help with providing measuring data. This paper was prepared by the cooperation of research teams "BME-RAIL" and "SZE-RAIL". The authors also thank the technical support of J. Lógó, G. Pintyőke, D.B. Merczel, and G. Kormos.

Conflicts of Interest: The authors declare no conflict of interest.

Nomenclature

Symbols	Description (Meaning)	Units
$\rho(x)$	radius of the elastic beam at the abscissa (x)	m
$\overline{x}_{TG,i}$	the track gauge value at the i-th measurement point	mm
$\overline{x}_{TG}$	the average of the individual values of the track gauge	mm
A	cross-sectional area of the rail	mm^2
C	bedding modulus (bedding factor) below the sleeper's bottom	N/mm^3
E	modulus of elasticity of the rail steel, E = 215,000 N/mm^2	N/mm^2
F_{comp}	longitudinal compression force	kN
F_{cr}	critical force inducing buckling of the track	kN
F_H	horizontal lateral force of support acting on the rail	kN
I	moment of inertia of the cross-section of the rail	mm^4
I_y	lower moment of inertia of the cross-section of the rail	mm^4
k	sleeper spacing	m
l_{dist}	wavelength of the distortion	mm
L_{lrb}	length of the longitudinal replacement beam	mm
l_{ws}	length of the rail without lateral support	m
$M(x)$	bending moment generated in the beam due to initial geometrical distortion	Nmm
Q	axle load of the vehicle	kN
s	width of the longitudinal substitutional beam	mm
S_{TG}	the value of the cumulative deviation from the mean at a given measurement point 'i'	mm
t	factor depending on the required safety	—
V	speed	km/h
x	*x-direction* based on Figures 6 and 7	-
x_{dist}	absolute distance from the cross-section	mm
xx	*xx-direction* based on Figures 6 and 7	-
y	*y-direction* based on Figures 6 and 7	-
Y	horizontal lateral force	kN
y_k	initial geometrical distortion	mm
$y_{k,max}$	maximum value of the initial geometrical distortion	mm
Y_{max}	maximum value of horizontal lateral force	kN
$Y_{max,sleeper}$	maximum value of horizontal lateral force acting on one sleeper	kN
yy	*yy-direction* based on Figures 6 and 7	-
Z	wheel load of the vehicle	kN
z	*z-direction* based on Figures 6 and 7	-
zz	*zz-direction* based on Figures 6 and 7	-
α	coefficient of linear thermal expansion of the rail steel, $\alpha = 1.15 \times 10^{-5}$ 1/°C	1/°C
β	dynamic multiplication factor	—
δ	factor depending on the quality of the superstructure	—
ΔT	deviation of the temperature of the rail from the actual neutral temperature	°C
η	value of the influence line of deflection	—
λ	coefficient dependent on the fixation of the end of the beam	—
σ_f	yield strength of rail steel	N/mm^2
$\tau(x)$	angle of the tangent drawn to the geometry of the structure	—
φ	speed-dependant factor	—

Abbreviations

Abbreviation	Meaning
ANN	artificial neural network
CC	circular curve
CWR	continuously welded rail
EC	entire curve
EMD	Empirical Mode Decomposition
FAS	Fourier Amplitude Spectrum
ICP	iterative closest point
IMF	Intrinsic Mode Function

IoT	Internet of Things
IoT-DTG	Internet of Things enhanced digital track gauge
LCD	liquid crystal display
LED	light emitting diode
MÁV CRTI Ltd.	MÁV Central Rail and Track Inspection Ltd., Budapest, Hungary
MÁV or MÁV Ltd.	Hungarian State Railways Ltd., Budapest, Hungary
MGT	million gross tons
MSE	mean squared error
R&D	research and development
SVM	support vector regression
TC	transition curve
TEN-T	Trans-European Transport Network
TG	track gauge
TQI	Track Quality Index
UIC	International Union of Railways
VMD	Variational Mode Decomposition
WiFi	wireless fidelity

References

1. Czére, B. *A Vasút Története (The History of Railway—In Hungarian)*; Corvina Kiadó: Budapest, Hungary, 1989.
2. Nøland, J.K. Hydrogen Electric Airplanes: A Disruptive Technological Path to Clean up the Aviation Sector. *IEEE Electrif. Mag.* **2021**, *9*, 92–102. [CrossRef]
3. Atilhan, S.; Park, S.; El-Halwagi, M.M.; Atilhan, M.; Moore, M.; Nielsen, R.B. Green Hydrogen as an Alternative Fuel for the Shipping Industry. *Curr. Opin. Chem. Eng.* **2021**, *31*, 100668. [CrossRef]
4. Gáspár, L. Lifetime Engineering Principles and Durable Roads. *Int. J. Pavement Eng. Asph. Technol.* **2016**, *17*, 58–72.
5. Gáspár, L.; Bencze, Z. Increasing Life Expectancy of Road Pavements. *Gradjevinar* **2020**, *72*, 515–522. [CrossRef]
6. He, L.; Cai, H.; Huang, Y.; Ma, Y.; Van Den Bergh, W.; Gaspar, L.; Valentin, J.; Vasiliev, Y.E.; Kowalski, K.J.; Zhang, J. Research on the Properties of Rubber Concrete Containing Surface-Modified Rubber Powders. *J. Build. Eng.* **2021**, *35*, 101991. [CrossRef]
7. Szpotowicz, R.; Tóth, C. Revision of Sustainable Road Rating Systems: Selection of the Best Suited System for Hungarian Road Construction Using Topsis Method. *Sustainability* **2020**, *12*, 8884. [CrossRef]
8. Primusz, P.; Tóth, C. Use of the Modified Ramberg-Osgood Material Model to Predict Dynamic Modulus Master Curves of Asphalt Mixtures. *Materials* **2023**, *16*, 531. [CrossRef]
9. Fischer, S.; Harangozó, D.; Németh, D.; Kocsis, B.; Sysyn, M.; Kurhan, D.; Brautigam, A. Investigation of Heat-Affected Zones of Thermite Rail Weldings. *Facta Univ. Ser. Mech. Eng.* 2023. [CrossRef]
10. Fischer, S.; Németh, D.; Horváth, H. Investigation of the Track Gauge in Curved Sections, Considering Hungarian Railway Lines. *Infrastructures* **2023**, *8*, 69. [CrossRef]
11. Fischer, S.; Liegner, N.; Bocz, P.; Vinkó, Á.; Terdik, G. Investigation of Track Gauge and Alignment Parameters of Ballasted Railway Tracks Based on Real Measurements Using. *Infrastructures* **2023**, *8*, 26. [CrossRef]
12. Németh, D.; Horváth, H.; Rad Movahedi, M.; Németh, A.; Fischer, S. Investigation of the Track Gauge in Straight Sections, Considering Hungarian Railway Lines. *Acta Polytech. Hung.* **2022**, *19*, 155–156. [CrossRef]
13. Nugymanova, G.; Nurgaliyeva, M.; Zhanbirov, Z.; Naumov, V.; Taran, I. Choosing a Servicing Company's Strategy While Interacting with Freight Owners at the Road Transport Market. *Sci. Bull. Natl. Min. Univ.* **2021**, *1*, 204–210. [CrossRef]
14. Ramazan, B.; Mussaliyeva, R.; Bitileuova, Z.; Naumov, V.; Taran, I. Choosing the Logistics Chain Structure for Deliveries of Bulk Loads: Case Study of the Republic Kazakhstan. *Natsional'nyi Hirnychyi Universytet. Nauk. Visnyk* **2021**, *18*, 142–147. [CrossRef]
15. Macura, D.; Laketić, M.; Pamučar, D.; Marinković, D. Risk Analysis Model with Interval Type-2 Fuzzy FMEA—Case Study of Railway Infrastructure Projects in the Republic of Serbia. *Acta Polytech. Hung.* **2022**, *19*, 103–118. [CrossRef]
16. Weinberg, S.J.; Sanches, F.; Ide, T.; Kamiya, K.; Correll, R. Supply Chain Logistics with Quantum and Classical Annealing Algorithms. *Sci. Rep.* **2022**, *13*, 4770. [CrossRef] [PubMed]
17. Yingfei, Y.; Mengze, Z.; Zeyu, L.; Ki-Hyung, B.; Andriandafiarisoa Ralison Ny Avotra, A.; Nawaz, A. Green Logistics Performance and Infrastructure on Service Trade and Environment-Measuring Firm's Performance and Service Quality. *J. King Saud. Univ. Sci.* **2022**, *34*, 101683. [CrossRef]
18. *EN 13848-1: Railway Applications—Track—Track Geometry Quality—Part 1: Characterisation of Track Geometry*; CEN European Committee for Standardization: Brussels, Belgium, 2008.
19. Chandra, S.; Agarwal, M.M. *Railway Engineering*, 2nd ed.; Oxford University Press: New Delhi, India, 2015.
20. Tzanakakis, K. History of the Track Gauge. In *The Railway Track and Its Long Term Behaviour*; Springer Tracts on Transportation and Traffic; Springer: Berlin/Heidelberg, Germany, 2013; Volume 2, ISBN 9783642360503.
21. Monfort, A. The Track Gauge in Spain: The Endless History. *Rev. Obras Publicas* **2019**, *166*, 64–75.
22. Puffert, D.J. The Standardization of Track Gauge on North American Railways, 1830–1890. *J. Econ. Hist.* **2000**, *60*, 933–960. [CrossRef]

23. Puffert, D.J. Path Dependence in Spatial Networks: The Standardization of Railway Track Gauge. *Explor. Econ. Hist.* **2002**, *39*, 282–314. [CrossRef]
24. Wang, C.; Li, X.; Chen, P.; Xie, Y.; Liu, W. Spatial Pattern and Developing Mechanism of Railway Geo-Systems Based on Track Gauge: A Case Study of Eurasia. *J. Geogr. Sci.* **2020**, *30*, 1283–1306. [CrossRef]
25. Zheng, S.; Peng, L.; Zhong, Q.; Li, L. Gauge and Wear Detection Method for Portable Track Inspection Trolley. *Zhendong Ceshi Yu Zhenduan/J. Vib. Meas. Diagn.* **2022**, *42*, 600–605. [CrossRef]
26. Pay, Z.H.; Ng, A.K.; Zhang, Z.A. Design and Fabrication of an Internet of Things Enhanced Digital Track Gauge. In Proceedings of the 8th International Conference on Control, Automation and Robotics, ICCAR 2022, Xiamen, China, 8–10 April 2022; IEEE: New York, NY, USA, 2022; pp. 45–48.
27. Yilmazer, M.; Karakose, M.; Aydin, I. Determination of Railway Track Gauge with Image Processing. In Proceedings of the 2021 International Conference on Data Analytics for Business and Industry, ICDABI 2021, Virtual, Online, 25–26 October 2021; IEEE: New York, NY, USA, 2021; pp. 510–514.
28. Zhang, G.; Ma, Z.; Yuan, J.; Kang, D.; Yan, D.; Li, J. Track Gauge Measurement Based on Wheel-Rail Lateral Relative Displacement. *Guangdian Gongcheng/Opto-Electron. Eng.* **2020**, *47*, 190252. [CrossRef]
29. Shi, H.; Xu, M.; Yu, Z. Track Gauge Measurement Method Based on Least-Square Curve Fitting Theory. *Tiedao Xuebao/J. China Railw. Soc.* **2019**, *41*, 81–88. [CrossRef]
30. Tsubokawa, Y.; Ishikawa, T. Development of a Dynamic Track Measuring Device for Gauge and Twist to Reduce Derailment Accidents. *WIT Trans. Built Environ.* **2018**, *181*, 253–262. [CrossRef]
31. Tang, W.B.; Long, Y.H.; Xie, Y.J.; Li, H.J.; Tang, C.B.; Zhang, Z.X. A Channel Rail Gauge Detection System of Modern Tram. In Proceedings of the 15th IEEE International Conference on Networking, Sensing and Control (ICNSC 2018), Zhuhai, China, 27–39 March 2018; IEEE: New York, NY, USA, 2018; pp. 1–5.
32. Aharkov, O.V.; Tverdomed, V.M.; Boiko, V.D.; Kovalchuk, V.V.; Strelko, O.H. Influence of the Structural Design of Rail Fastenings on Ensuring the Stability of Track Gauge in Operating Conditions. In Proceedings of the IOP Conference Series: Materials Science and Engineering—8th International Scientific Conference on Reliability and Durability of Railway Transport Engineering Structures and Buildings, TRANSBUD 2019, Kharkiv, Ukraine, 20–22 November 2019; Volume 708, p. 012001.
33. Ahac, M.; Lakusic, S. Tram Track Maintenance-Planning by Gauge Degradation Modelling. *Transport* **2015**, *30*, 430–436. [CrossRef]
34. Akkerman, G.; Akkerman, S. Rail Gauge Trajectory, Macro and Micro Profile of Tracks. In Proceedings of the IOP Conference Series: Materials Science and Engineering—8th International Scientific Conference Transport of Siberia 2020, Novosibirsk, Russia, 22–27 May 2020; Volume 918, p. 012153.
35. Falamarzi, A.; Moridpour, S.; Nazem, M.; Cheraghi, S. Prediction of Tram Track Gauge Deviation Using Artificial Neural Network and Support Vector Regression. *Aust. J. Civ. Eng.* **2019**, *17*, 63–71. [CrossRef]
36. Hungarian State Railways D.54, Construction and Track Maintenance Technical Data, Regulations. 1988. Available online: https://www.mavcsoport.hu/mav-csoport/bemutatkozas/szabvanykozpont (accessed on 2 June 2023).
37. European Committee for Standardization EN 206:2013+A2:2021: Concrete—Specification, Performance, Production and Conformity. 2021, pp. 1–103. Available online: https://genorma.com/en/project/show/cen:proj:73198 (accessed on 2 June 2023).
38. European Committee for Standardization EN 10138-2:2005: Prestressing Steels—Part 2: Wire. 2005, pp. 1–15. Available online: https://global.ihs.com/doc_detail.cfm?item_s_key=00496117 (accessed on 2 June 2023).
39. Esveld, C. *Modern Railway Track*; MRT Production: Zaltbommel, The Netherlands, 2001; ISBN 978-1-326-05172-3.
40. Lichtberger, B. *Track Compendium*; Eurailpress Tetzlaff-Hestra GmbH & Co. Publ.: Hamburg, Germany, 2005.
41. Fischer, S.; Eller, B.; Kada, Z.; Németh, A. *Railway Construction*; Universitas-Győr Nonprofit Kft.: Gyor, Hungary, 2015; ISBN 978-615-5298-69-1.
42. European Committee for Standardization EN 1992-1-1:2004/A1:2014: Design of Concrete Structures—Part 1-1: General Rules and Rules for Buildings. 2014, pp. 1–226. Available online: https://infostore.saiglobal.com/en-us/standards/en-1992-1-1-2004-a1-2014-333210_saig_cen_cen_765732/ (accessed on 2 June 2023).
43. Hungarian State Railways D.12/H, Instruction for the Construction, Maintenance and Supervision of CWR Tracks. 2009, pp. 1–123. Available online: https://www.mavcsoport.hu/mav-csoport/bemutatkozas/szabvanykozpont (accessed on 2 June 2023).

vehicles

MDPI

Review

Challenges in the (Re-)Connection of Peripheral Areas to the Rail Network from a Rolling Stock Perspective: The Case of Germany

Benedikt Hertel [1,*], Johannes Pagenkopf [1] and Jens König [2]

[1] Institute of Vehicle Concepts, German Aerospace Center (DLR), 12489 Berlin, Germany; johannes.pagenkopf@dlr.de
[2] Institute of Vehicle Concepts, German Aerospace Center (DLR), 70569 Stuttgart, Germany; jens.koenig@dlr.de
* Correspondence: benedikt.hertel@dlr.de

Abstract: Currently, in the course of the German mobility transition, an increasing number of disused rail lines are already being or intended to be reactivated in order to increase capacities, decrease transport-related emissions and reconnect rural areas to passenger rail services, thus creating a more comprehensive rail service. However, the use of state-of-the-art regional railcars on old rural infrastructure often leads to problems since they are often worn out and do not meet today's technical standards. This applies, for example, to the axle loads and dimensions of the vehicles but also to operational aspects, such as the vehicle's passenger capacity and accessibility. First, this work gives an overview of the available rolling stock and the given infrastructure, as well as an analysis of the (system) interfaces. Subsequently, various challenges for the re-connection of peripheral areas to the rail network were identified through data research and comparison of the vehicle and infrastructure parameters. In addition, the requirements related to possible autonomous operation and the related absence of the driver and crew personnel in the vehicle, which require new solutions in terms of safety, were taken into consideration. Orientation of future rolling stock generations towards the existing infrastructure and the required transport needs, including lower axle loads, accessibility and smaller capacities, can contribute to the economic operation of low-capacity lines and bring more passengers to public transport.

Keywords: rolling stock; rail infrastructure; reactivated rail lines; regional rail; branch line; accessibility

1. Introduction

The aim to shift the current transport system is accompanied by the need for new types of rail vehicle concepts. In rural areas, in particular, there are low-frequency and abandoned rail lines on which the operation of large and heavy railcars on a scheduled basis is not economically viable. However, the development of formerly closed or decommissioned branch lines has positive effects on the attractiveness of the regions. A reconnection to the rail network can represent, among other benefits, a locational advantage for the reconnected areas [1,2]. However, in order to create an attractive regional rail transport, it is also necessary to have a reliable system with good connections and service frequencies [3].

Since branch lines with the potential for reactivation are generally located in sparsely populated areas, vehicles with low capacities are needed to serve them. At the same time, vehicles are developing into modular, multiple-unit vehicle families that are intended to cover the widest possible range of applications. Due to high maximum speeds, crash and comfort requirements, as well as large passenger capacities, the weight and size of the vehicles have increased significantly since the introduction of rail buses in the 1930s to 1960s.

Current research mostly focuses on dedicated technical or operational problems in the general operation of railroads in rural areas, such as vehicle weight [4,5], alternative powertrains [6] or driverless operation [7,8]. The aim of this article is not to deepen individual technical and operational issues but to draw a holistic picture of the challenges involved in operating low-frequency branch lines and reactivating old lines, considering the various interfaces between rolling stock, infrastructure and passengers. For this, a systemic approach was chosen to elaborate on the challenges that arise when reactivating low-frequency rail lines and to illustrate the requirements that must be met by future generations of rail vehicles operating in rural areas.

In order to find out whether and which gaps exist in the current rolling stock portfolio for the operation of low-frequency routes, especially in peripheral areas, rolling stock currently in operation and market-available rail vehicles were first examined and compared with the infrastructural and capacity demand. The state-of-the-art vehicle research focused on single- and two-car units that are currently being used or have been used in the past on branch lines. Components of the analysis include the development of capacity, car body design and design speed over time and vehicle generations. In addition, various concept studies and demonstrators are included in the evaluation. On the infrastructure side, potential future fields of operation were identified. The focus is on lines where the use of market-available vehicles would most probably not be economically viable. Use cases were formed for the lines under consideration on the basis of reactivation studies and further technical data. Based on this, potential conflicts between the vehicle and the existing infrastructure were identified. Finally, based on the initial situation, the challenges that need to be considered in the development of future rolling stock for peripheral areas are highlighted.

2. Initial Position

2.1. Evolution of Rolling Stock in Branch Line Service

In the previous decades, there have been major changes in the field of rail rolling stock for branch line service, from light rail busses in the 1950s to heavy multiple units today. In the 1950s and 1960s, various versions of light rail buses—such as the VT 95/98 (Deutsche Bundesbahn in the Federal Republic of Germany (FRG)) or VT 2.09 (Deutsche Reichsbahn in the former German Democratic Republic (GDR))—were developed for use on branch lines. These small two-axle vehicles usually had about 60 seats. Due to their lightweight design dimensioned for a longitudinal compressive force of 500 kN at buffer level, they were very light vehicles (approx. 14–20 t, depending on the specific vehicle) [9]. Except for a few examples in museum operation, these vehicles are now retired.

In the mid-1970s, the single- and two-car DB class 627 and class 628 were introduced in Germany as replacements for the rail buses. As a result of serious accidents involving first-generation rail buses with heavy rail vehicles, the required longitudinal compressive force at the buffer level for new railcars was increased to 1500 kN. The increased safety requirements were reflected in a significant increase in vehicle masses (e.g., VT 95 with approx. 14 t and class 627 with approx. 34 t with comparable passenger capacity) [9].

A second generation of lightweight rail buses in Germany was developed in the mid-1990s. Two new vehicles, the DWA LVT/S and the DWA double-decker railcar, were introduced in 1996. Both vehicles had a low longitudinal compressive force of 600 kN or 400 kN, respectively. The reduced passive safety (lower longitudinal compressive forces) was compensated for by an increased active safety—among other things, through better braking technology. While some of the LVT/S are still in service with private operators today [10], the double-decker railcar is mostly withdrawn from regular service due to susceptibility to faults, among other reasons. Only a small number of both vehicles were manufactured [11,12].

Simultaneously with the lightweight vehicles, vehicles such as the Regio-Shuttle RS1, the class 650 from Stadler (formerly Adtranz) or the LINT 27 (class 640) out of the Coradia LINT family of Alstom entered the market, which in the long term have become established above all for branch lines. These vehicles are of conventional design with a longitudinal compressive force of 1500 kN. Compared with the first- and second-generation rail buses, they show higher specific masses per seat as well as higher axle loads than the rail buses [11].

Newer vehicle types used in regional and commuter services are generally derived from modular multi-unit train platforms. They are designed for higher speeds of up to 140 or 160 km/h. In order to be able to operate them on main lines and TEN-lines, they are equipped with high frame rigidity, just like the Regio-Shuttle RS1 or the Alstom Coradia LINT family. Traction units are usually classified in category P-II of the standard DIN EN 12663-1 [13] (TSI conform) and must resist a longitudinal compressive force on a buffer level of 1500 kN. DIN EN 15227 [14], issued in 2008, regulates the crashworthiness design of vehicles.

The further increase in safety requirements resulting from crashworthiness adapted design and vehicle designs for higher speeds (cf. Figure 1) has also led to a further increase in vehicle masses. Furthermore, the number of seats increased to up to 120 per vehicle.

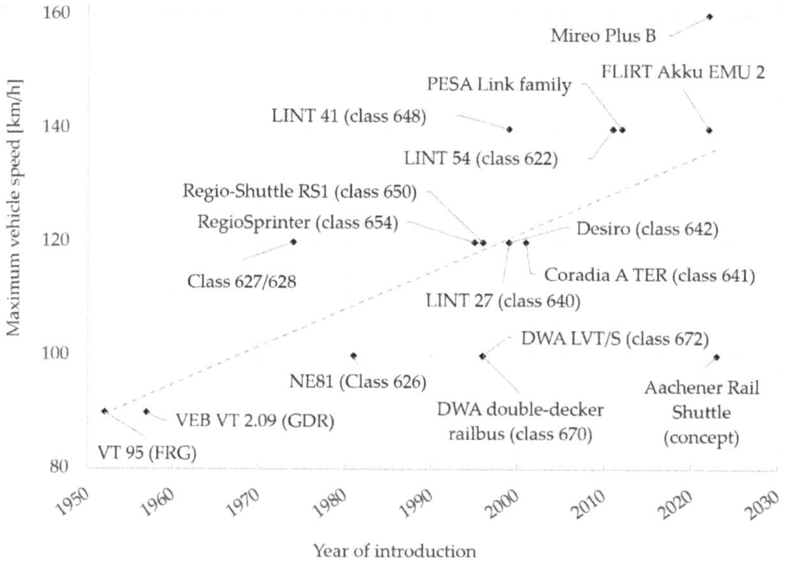

Figure 1. Development of maximum vehicle speeds over time (Reference: various publicly available information from OEM and other sources).

Figure 2 shows the development of axle loads and seating capacities of regional rail vehicles over time. It can be seen that the seating capacities, as well as the axle loads of the vehicles, have increased over time. Vehicles for branch lines had maximum axle loads of between 10 and 14 t until the 1980s. This allowed them to be used on all current line categories. With 14 t up to 18 t, vehicles developed in the 1990s—such as the Stadler Regio-Shuttle RS1 (14 to 15 t depending on the series) or Alstom LINT 41 (18 t)—have a significantly higher mass. Vehicles with an axle load of over 16 t can only be operated on class B lines or higher. Current multiple units have maximum axle loads of between 18 t (diesel multiple units such as the Pesa Link II) and 20 t (battery electric vehicles such as the Mireo Plus B from Siemens), depending on the type of traction. Due to the high axle loads, operation on some branch lines is no longer possible. Other reasons for an increase in weight include changes in comfort requirements and additional HVAC and passenger information systems.

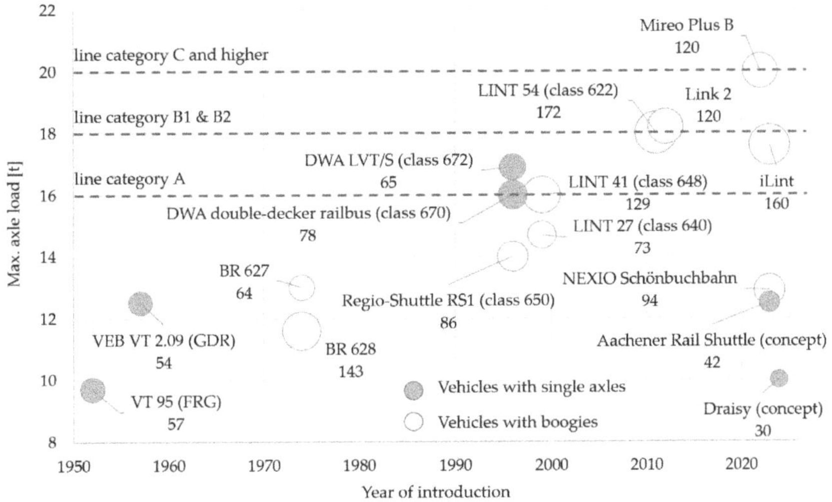

Figure 2. Development of axle loads and seating capacities (bubble size and number) of regional rail vehicles in Germany; variations due to different vehicle series possible (Reference: various publicly available information from OEM and other sources).

Various lightweight vehicle concepts for different applications are currently being developed throughout Europe. In addition to the Coventry Very Light Rail (VLR), which is being developed as a novel light rail concept, activities are primarily focused on the development of small vehicles to maintain operations on low-demand lines and reconnect rural areas to the rail network. One example is the *Aachener Rail Shuttle* (ARS), which is being developed at the Institute for Rail Vehicles and Transport Systems (RWTH Aachen). The ARS is an autonomous, battery electric vehicle (concept) in lightweight construction with a comparatively low passenger capacity of 90 passengers (seated and standing) and a maximum speed of 100 km/h [15,16].

Vehicle concepts such as the Taxi Rail in France are vehicles that are specifically oriented to the conditions of routes to be reactivated [16]. This is in line with the plans to reconnect peripheral areas to the rail network and to shift more traffic from road to rail. France's national railway company, SNCF, is currently working on three different vehicle concepts with its partners. Flexy, Draisy and Innovative Light Train each represent individual concepts with capacities ranging from nine to eighty seats. All of them are designed to revitalize rural areas [17].

The class 455 electric multiple unit for the Schönbuchbahn is already on the rails but has not yet been certified. In order to replace the existing Stadler Regio-Shuttle RS1, a vehicle procurement tender was issued in 2013. However, all bids showed significant deviations from the requirements (vehicle weight, energy demand and capacity in relation to the vehicle length). Due to that, the Spanish manufacturer CAF was awarded the contract with a tram-train approach based on the NEXIO platform. The vehicle, which was originally planned according to LNT guidelines, has an axle load of only 12.3 t despite being fully loaded, which is a significantly lower axle load compared to other currently available traction unit families [4].

Since class 455 cannot be approved for every line according to the LNT guideline, and instead only according to the EBO, the braking system now has excessively high deceleration values, which is not allowed according to the EBO but would be necessary to fulfill the LNT requirements with the reduced carbody loads compared to the usual trains. The approval of the vehicles is now to be route-specific [18].

2.2. Potential Areas of Application and Derivation of a Use Case

2.2.1. Data Base

In order to meet the goal set out in the current German coalition agreement from 2021 of doubling local passenger transport services by 2030 and reactivating closed lines, among other things, an increasing number of disused lines are being reactivated, and operations resumed [19]. In addition, efforts are being made to operate existing lines, which were previously served by diesel vehicles, with locally emission-free vehicles in the future. Based on this situation and the state-of-the-art rolling stock described, the question arises which routes can be served with the available vehicles and for which new concepts are required. For this purpose, the infrastructural and operational framework conditions of the routes being considered were first described in a use case.

In order to create the use case on specific deployment scenarios, the use case was created on the basis of routes foreseen for reactivation. In addition to the infrastructure characteristics, the expected passenger potential on the lines is one of the main factors determining the requirements for such vehicles.

In order to be able to make a prediction about the required passenger capacities and, therefore, about vehicle sizes, studies of the potential for reactivating disused lines were used. A study with a comprehensive database is the "*Potenzialanalyse zur Reaktivierung von Schienenstrecken in Baden-Württemberg*" [20]. In addition to route information, this study analyzes the respective passenger potential of the individual routes, broken down by route sections between stops. In order to estimate the passenger potential, the study considered the catchment area of the route and used this to produce, among other things, a forecast of the number of passengers boarding the train and the maximum cross-sectional load for each route section. For further consideration, the selection of routes available from the study was further narrowed down. Routes that could potentially be served by modern, available vehicles were not included in the analysis.

The study focuses, in particular, on the potential passenger volumes as well as operational parameters. This is important in order to make a statement on the worthiness of reactivation but does not provide a sufficient basis for considering the interfaces of the vehicles and infrastructure for compatibility. Therefore, further data on line and station infrastructure, as well as current or planned operations, were gathered. The data were compiled from documents published by the rail infrastructure companies in charge, such as the *Collection of Operational Regulations* (*Sammlung betrieblicher Vorschriften*) (SbV) or the *Network Statement* (*Schienennetz-Benutzungsbedingungen*) (SNB), which can be accessed on the "*Schienennetz-Nutzungsbedingungen (SNB)-Portal*" of the Federal Network Agency (*Bundesnetzagentur*) [21].

2.2.2. Vehicle Interfaces with the Surroundings

To create the use cases, the interfaces between the rail vehicles and the elements of the "rail system" are regarded. These are in particular:

- **Track infrastructure:** especially superstructure and substructure, including the track system, elevated and underground structures such as bridges or tunnels, as well as level crossings and signaling systems.
- **Station infrastructure:** The relevant parameters are platform height and length, with the former being decisive in terms of accessibility.
- **Power supply:** Classically by continuous contact wire or diesel at the filling station, and also via electrants, catenary island systems, charging stations or hydrogen filling stations in the future.
- **Train control and signaling systems (TCSS):** Depending on the line, the equipment varies widely, from deconstructed systems to modern ECTS systems. On less frequented secondary lines, only simple equipment such as PZB tends to be available.
- **Passengers:** In order to create acceptance of the vehicle, certain minimum requirements must be met in various areas, such as safety (security), accessibility, service frequency and comfort.

2.2.3. Derivation of the Use Case

The passenger and infrastructure data were the base for the creation of a generic use case, which is intended to represent a cross-section of potential fields of application (set of characteristic routes). From the pool of data, the governing parameters were chosen so that the use case would include as many application fields as possible. However, outlier values were not included to avoid stretching the technical requirements too far.

The use case covers low-frequency routes in rural areas for which current vehicles have too-high passenger capacities or too-high mass. A connection to the higher-level regional rail network is guaranteed at one or both ends of the line. The technical parameters of the line and the stations, as well as the transportation requirements, are listed in Table 1. The parameters result from an analysis of different sources, including [20], as well as an analysis of the corresponding SbV and SNB of different railroad infrastructure companies [21].

Table 1. Use Case Parameters.

Trackside Infrastructure Parameters of the Use Case		
Importance of the line		Secondary/branch line
Max. track length	[km]	30
Distance betw. stops (min.-max.)	[km]	2.4 (1.2–3.4)
Electrification		no
Number of tracks		Singe track
Speed limit	[km/h]	60 (80)
Line category		A
Minimum curve radius	[m]	140
Max. gradient	[‰]	32
Loading gauge		EBO G1
Station infrastructure		
Platform height		Wide range of platform heights (0–550 mm above rail level)
Platform length		Partially less than 30 m
Service requirements		
Capacity per hour and direction		10–60 (peak time)
Travel time (average/max)	[min]	12/40
Extended mobility needs		Wheelchairs/rollators, strollers, (e-)bikes, luggage, etc.

Service (transport) demand was approximated by the maximum cross-section volume, assuming a load ratio of 3:1 from peak to off-peak hours. The determined required capacity at rush hour serves as a guideline for determining the required passenger capacity. The travel time was roughly determined using a constant travel scenario with constant acceleration and deceleration. Energy saving driving modes would increase the minimum travel time.

On some of the lines studied, there is currently no traffic, while on other lines, there is freight traffic and/or occasional passenger traffic (including seasonal traffic). The equipment standard of the TCSS and the operating procedures used vary greatly. In some cases, the TCSS have already been completely dismantled.

3. Development Premises for Rail Vehicles on Reactivated Lines and Low-Frequented Branch Lines and Discussion of the Results

Based on the use case, various challenges and requirements for future rail vehicles can be derived. These mostly result from the infrastructural conditions. Many of the examined lines have not been adapted to the current state-of-the-art design for years or decades. As a result, some of the track systems and stations are in a heavily antiquated condition.

Besides the technical challenges, however, it is also important to consider the economic factors. From the 1950s to the 1990s—and in some cases even beyond—numerous passenger services were discontinued on branch lines with low-capacity utilization. This was usually justified by excessive costs, and the service was replaced by bus transportation. However, bus services alone do not offer the same transport quality, comfort, etc., as rail services. A connection of a locality to the rail network can contribute to the upgrading of a region and influence the choice of a place of residence. Therefore, an attractive public transport system requires an integrated concept of bus and rail services that complement each other [1].

3.1. Accessibility

With a view to future transport services in rural areas, it is important to ensure an attractive and non-discriminatory public transport mobility for passengers in all life situations. This also results in technical challenges for the design of rail vehicles. An elementary aspect of implementation is an overall system that is as accessible as possible, including not only the vehicle itself but also the transitions between the transport stations and the next traffic medium. Particularly, rail transportation offers advantages over other modes of transport, such as buses, due to the greater space available in the vehicle or the fact that vehicles often have level floors.

To enable accessible boarding of the vehicle, the floor and platform must be kept at the same level (as well as the distance from the vehicle to the platform as small as possible). However, platform heights vary widely in Germany, especially in rural areas. While the traffic stations of highly frequented lines have standardized platform heights of 550 mm or 760 mm, according to TSI, and are thus compatible with many of the modern commuter trains, platforms at stations and stops in rural areas often still have heights of 0 to 380 mm above rail level. In the long term, as part of the standardization process, platforms are expected to be adjusted to 550 mm or 760 mm *. However, these changes will take some time and will take place on lines that are more heavily used first. Therefore, solutions must be found that will allow accessible use of all transport stations even in the short term. Currently, mostly mobile ramps, among other things, are used to overcome small changes in height from the platform to the vehicle. However, in practice, these do not allow independent use of public transport, and although they comply with the regulations due to gradients of up to 18 percent (TSI PRM [22]), they are a serious hurdle to using public transport for many people [23].

* The current target platform heights from the platform height concept (*Bahnsteighöhenkonzept*) of Deutsche Bahn AG (DB AG) at the federal level and those of the states vary. DB AG aims to standardize all platforms to a height of 760 mm. Many federal states, on the other hand, have recently raised platforms to 550 mm or are in the process of doing so since this is in line with the boarding heights of most modern regional rail vehicles, especially double-decker cars [24].

The boarding heights of present vehicles have heights between 550 mm and 800 mm (see Figure 3). This is consistent with the standardization concept for platform facilities but does not provide accessibility options at many of the existing infrastructures.

Also, the vehicle interior design must be oriented to the mobility needs in different phases of life and offer sufficient space. This requires, for example, sufficiently dimensioned and easily accessible multi-purpose areas that offer sufficient parking and securing options for baby carriages, wheelchairs, walkers, as well as (e-)bikes and luggage. The TSI PRM provides guidance for implementation [22].

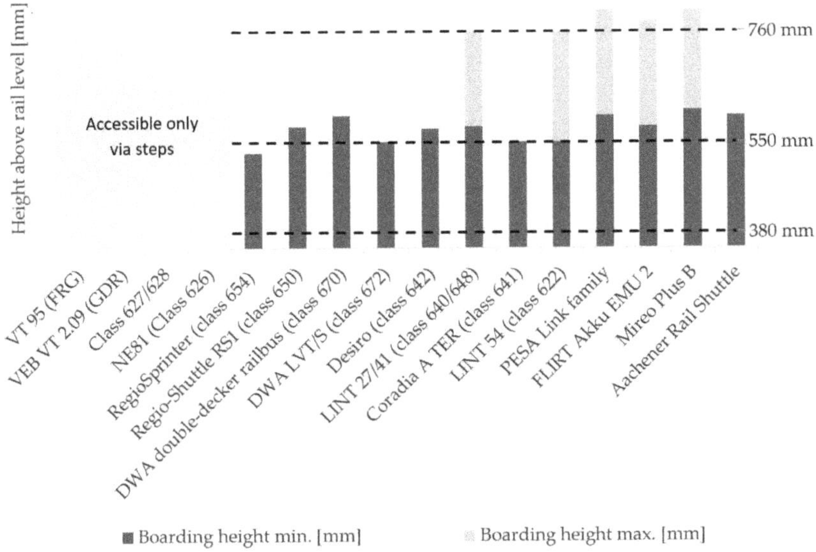

Figure 3. Boarding heights of different regional passenger transport vehicles used in Germany (floor height in the door area) (Reference: various publicly available information from OEM and other sources).

3.2. Driverless Driving

One way to reduce operating costs and provide a solution to staff shortage is to operate driverless, which is a significant issue, especially for vehicles with a smaller passenger capacity than the usual existing one. However, this raises new technical and operational challenges at various levels. On the technical level, solutions must be found to operate the vehicle safely over an unsecured guideway. Existing driverless vehicle systems have so far operated on secured routes, such as subway systems, for example, in the German city of Nuremberg. The route of a regional rail vehicle, on the other hand, has additional points of contact with other modes of transport, e.g., at level crossings, but also an increased risk of objects on the tracks, such as fallen trees, individuals or animals [25].

Driverless driving of vehicles in mass transit is the subject of various development and demonstration projects. Using "driving on sight," the Aachener Rail Shuttle of the Institute of Rail Vehicles and Transportation Systems (RWTH Aachen University) is also able to operate outside secured tracks. Research has been carried out, among other things, to determine the influence of track parameters, such as curve radius and loading gauge, as well as braking performance on safe travel speeds [7].

In addition to the technical safety of the rail vehicle (safety), the security of passengers inside the vehicle in relation to other passengers (in-vehicle security) must also be considered. Passengers consider in-vehicle security to be a very important factor when choosing a mode of transport. Even today, in vehicles that have driving and service personnel, the fear of being subject to attacks has a decisive influence on the choice of mode choice. The complete absence of personnel in the vehicle further increases the actual as well as perceived insecurity. Other concerns about driverless operations include increased pollution and vandalism, as well as a lack of support options. In order to create sufficient acceptance of the vehicle, it is necessary to address these issues and create solutions that restore safety as well as educate passengers [26,27].

3.3. Train Meets

Many reactivation projects or low-frequency lines are single-track lines with no or few passing places. This considerably reduces the performance of the lines. If a traffic offer with higher frequency or even demand-responsive transport is aimed for, train meets will unavoidably occur. For these situations, technical or operational solutions must be found to create ways of taking evasive action.

3.4. Energy Supply

Usually, reactivation studies and routes proposed by transport associations for reactivation are non-electrified branch lines. In order to establish a locally emission-free operation, infrastructural measures are, therefore, required to supply the vehicles with energy. Due to the development of battery- and hydrogen-powered vehicles, electrification of the entire line is no longer necessary for locally emission-free operation. The energy supply for the operation of battery-powered vehicles can be provided in various ways, e.g., via charging points at stations or partial electrification of the line. For the operation of hydrogen-powered vehicles, hydrogen refueling stations are required [28].

3.5. Vehicle Homologation

Strict requirements apply to the approval and operation of rail vehicles and infrastructure to ensure safe operation. However, rigid regulations, standards and laws can also slow down innovative solutions on the railways. New types of vehicle and operating concepts, which are tailored to a specific purpose and deviate from current operating structures, can fall through the existing grid of regulations and standards. In addition to the rededication of rail lines to light rail lines, other measures for the realization of simplified rail vehicles are conceivable. An isolated operation, separated from heavy rail traffic, can allow room for new safety concepts adapted to the circumstances.

3.6. Vehicle Mass and Axle Load

The majority of rail lines in Germany are designed for high axle loads. With 86%, line class D4 (22.5 t wheelset load, 8 t/meter load) is the most widely represented in the DB Netz network (as of 2021) [29]. Many lines are also being upgraded for route category D4 in the course of reactivation projects in order to make the lines suitable for freight traffic. However, some lines are still classified as category A or B2. On these lines, the use of BEMU vehicles with wheelset loads up to 20 t is not possible.

Even though the operation of heavy commuter railcars such as the Siemens Mireo Plus B with a wheelset load of 20 t is possible on many lines, it does entail some disadvantages. With a heavy vehicle structure, the demand for traction energy increases. This is particularly noticeable on regional rail lines where frequent acceleration is required due to short stopping distances. In the case of BEMU or HEMU vehicles, which are to replace diesel railcars on branch lines in the future, this also increases the need for energy storage capacity in the form of batteries or hydrogen tanks, which further increases the vehicle weight. In addition to the energy requirement, the maintenance costs for the track superstructure also increase with increasing wheelset load [30].

From a technical and economic point of view, therefore, there are a number of reasons in favor of lighter vehicles. On the other hand, however, there are high requirements for the vehicle design in terms of car body strength and crashworthiness. The TSI LOC&PAS [31] regulates the application of EN 12663 and EN 15227 which defines the strength requirements and crashworthiness for car bodies depending on the vehicle category.

4. Conclusions

The identified challenges illustrate the tension between the large number of requirements for the vehicle and the pressure to be as cost-efficient as possible in terms of acquisition and operation.

The evolution from small and light vehicles designed for operation on branch lines to modular vehicle families, which are supposed to cover as many operational scenarios as possible, is creating a gap in the vehicles currently available on the market. The infrastructure and passenger requirements cannot be met with these vehicles due to, for example, their high axle loads, capacities and lack of accessibility to the existing infrastructure. In addition, there is the cost pressure, which is on the modes of transport in general, but especially on low-utilized services. The energy supply of the vehicles must also be rethought if local emission-free operation is the goal.

Technical innovations such as driverless operation can help realize cost-effective operation but raise new challenges in the area of safety and in-vehicle security. To date, there is no autonomous rail vehicle approved in Germany for unsecured rail operations, and the physical absence of staff on board can lead to further uncertainty among passengers.

In order to re-establish a widely spread rail transport system in rural areas, new vehicle concepts are needed that can be adapted to the requirements of the infrastructure and the mobility needs, as well as low costs in acquisition and operation. In combination with new operating concepts, coordinated with connecting transport modes such as buses, a widespread coverage of rail-bound public transport can also be created in rural areas.

The content of this paper primarily represents the situation in Germany. However, similar questions also arise in other (European) countries, such as the Czech Republic or France. Due to the historically individually evolving infrastructures of the respective countries and regions, the requirements may partly differ from the requirements elaborated in this paper. An individual analysis of the specific requirements of the respective regions is therefore necessary in future work.

DLR is currently working on a new vehicle concept. Using a systematic approach that considers the operating concept, infrastructural aspects and a demand-oriented design of the vehicle, the aim is to create a small, lightweight, optimized, multi-modular, automatic driving rail vehicle with local emission-free propulsion.

Author Contributions: Conceptualization, B.H.; methodology, B.H. and J.P.; validation, B.H., J.P. and J.K.; formal analysis, B.H.; investigation, B.H., J.P. and J.K.; data curation, B.H.; writing—original draft preparation, B.H.; writing—review and editing, J.P. and J.K.; visualization, B.H.; supervision, J.P.; All authors have read and agreed to the published version of the manuscript.

Funding: This work was funded by the Helmholtz Association of German Research Centres on behalf of the Federal Ministry for Economic Affairs and Climate Action. The work was carried out as part of the Project VMo4Orte within the framework of DLR basic funding.

Data Availability Statement: Data sharing is not applicable to this article.

Conflicts of Interest: The authors declare no conflict of interest.

References

1. Verband Deutscher Verkehrsunternehmen e.V. *Auf der Agenda: Reaktivierung von Eisenbahnstrecken*; Verband Deutscher Verkehrsunternehmen e.V.: Köln, Germany, 2022. Available online: https://www.vdv.de/vdv-reaktivierung-von-eisenbahnstrecken-2022-3.-auflage.pdfx (accessed on 10 July 2023).
2. Rohs, M.; Flore, G.; Lepski, M. *Räumliche Effekte Reaktivierter Schienenstrecken im Ländlichen Raum*; BBSR-Online-Publikation: Bonn, Germany, 2022. Available online: https://www.bbsr.bund.de/BBSR/DE/veroeffentlichungen/bbsr-online/2022/bbsr-online-27-2022-dl.pdf;jsessionid=B0F7E2C4CCB5D04E17536CAE0791FD52.live21322?__blob=publicationFile&v=4 (accessed on 8 June 2023).
3. Bundesnetzagentur. Ergebnisse der Endkundenbefragung 2020 im Schienenpersonenverkehr. 2020. Available online: https://www.bundesnetzagentur.de/SharedDocs/Downloads/DE/Sachgebiete/Eisenbahn/Unternehmen_Institutionen/Endkundenbefragung/TiefenbefragungSPV2020.pdf?__blob=publicationFile&v=3 (accessed on 23 January 2023).
4. Bitterberg, U. Leichter Regional—Elektro-Triebzug (LRET) für die Schönbuchbahn. *ETR* **2018**, *6*, 17–21.
5. Winnett, J.; Hoffrichter, A.; Iraklis, A.; McGordon, A.; Hughes, D.J.; Ridler, T.; Mallinson, N. Development of a very light rail vehicle. *Proc. Inst. Civ. Eng.-Transp.* **2017**, *4*, 231–242. [CrossRef]
6. Streuling, C.; Pagenkopf, J.; Schenker, M.; Lakeit, K. Techno-Economic Assessment of Battery Electric Trains and Recharging Infrastructure Alternatives Integrating Adjacent Renewable Energy Sources. *Sustainability* **2021**, *13*, 8234. [CrossRef]
7. von Stillfried, A.; Schindler, C. Fahren auf Sicht–Ein Betriebskonzept für den fahrerlosen Nahverkehr. *ETR* **2020**, *10*, 22–27.

8. Flamm, L.; Meirich, C.; Jäger, B. Die Umsetzung des automatisierten Bahnbetriebs zwischen Technik, Regelwerken und Wirtschaftlichkeit. *ETR* **2019**, *3*, 27–31.
9. Reinhardt, W. *Öffentlicher Personennahverkehr: Technik-Rechtliche und Betriebswirtschaftliche Grundlagen; mit 33 Tabellen*, 1st ed.; Vieweg + Teubner: Wiesbaden, Germany, 2012; ISBN 978-3-8348-1268-1.
10. Hanseatische Eisenbahn. Unsere Fahrzeugflotte. Available online: https://www.hanseatische-eisenbahn.de/fahrzeuge.html (accessed on 1 February 2023).
11. Hondius, H. Zehn Jahre Entwicklung einer neuen Generation Regio-Dieseltriebwagen: Übersicht über die neuen Fahrzeuge des SPNV in Dieselnetzen. *Der Nahverkehr* **2006**, *5*, 34–42.
12. MDR Sachsen-Anhalt. Keine Investoren: Insolvente Molinari-Fahrzeugtechnik in Dessau Wird Geschlossen. Available online: https://www.mdr.de/nachrichten/sachsen-anhalt/dessau/dessau-rosslau/waggonbau-molinari-schliessung-insolvenz-kuendigungen102.html (accessed on 13 July 2023).
13. *DIN EN 12663-1:2010*; DIN Deutsches Institut für Normung e.V. Bahnanwendungen—Festigkeitsanforderungen an Wagenkästen von Schienenfahrzeugen—Teil 1: Lokomotiven und Personenfahrzeuge (Und Alternatives Verfahren für Güterwagen). Beuth Verlag: Berlin, Germany, 2010.
14. *DIN EN 15227*; DIN Deutsches Institut für Normung e.V. Bahnanwendungen—Anforderungen an die Kollisionssicherheit von Schienenfahrzeugen. Beuth Verlag: Berlin, Germany, 2011.
15. Schindler, C. The Aachen Rail Shuttle ARS—Autonomous and energy self-sufficient feeder transport. *J. Rail Transp. Plan. Manag.* **2022**, *21*, 100299. [CrossRef]
16. Haydock, D.; Müller, C. Taxirail und VLR: Neue ultraleichte Fahrzeuge. *RailBusiness* **2021**, *20*, 6.
17. Christeller, R. Developing concepts to revive rural lines. *Railw. Gaz. Int.* **2023**, *179*, 22–25.
18. Müller, C. Zulassung nicht mehr nach LNT-Richtlinie. *RailBusiness* **2022**, *44*, 5.
19. SPD und Bündnis 90/Die Grünen und FDP. Mehr Fortschritt Wagen—Bündnis für Freiheit, Gerechtigkeit und Nachhaltigkeit—KOALITIONSVERTRAG 2021–2025. 2022. Available online: https://www.spd.de/fileadmin/Dokumente/Koalitionsvertrag/Koalitionsvertrag_202 (accessed on 31 July 2022).
20. PTV Transport Consult GmbH. Potenzialanalyse zur Reaktivierung von Schienenstrecken in Baden-Württemberg Vergleichende Relationsbezogene Ermittlung des Fahrgastpotenzials. 2020. Available online: https://vm.baden-wuerttemberg.de/fileadmin/redaktion/m-mvi/intern/Dateien/PDF/PM_Anhang/Potenzialanalyse_PTV_Bericht_01.pdf (accessed on 12 March 2023).
21. Bundesnetzagentur. Schienennetz-Nutzungsbedingungen (SNB)-Portal. Available online: https://www.bundesnetzagentur.de/DE/Fachthemen/Eisenbahnen/SNB-Portal/snb-portal-node.html (accessed on 11 December 2021).
22. COMMISSION REGULATION (EU) No 1300/2014 of 18 November 2014 on the Technical Specifications for Interoperability Relating to Accessibility of the Union's Rail System for Persons with Disabilities and Persons with Reduced Mobility: TSI PRM. 2014. Available online: https://eur-lex.europa.eu/legal-content/EN/TXT/PDF/?uri=CELEX:32014R1300&from=DE (accessed on 10 July 2023).
23. Geisslinger, E. Für Rollis Untauglich: Neue Züge in Schleswig-Holstein. Taz [Online]. Available online: https://taz.de/Neue-Zuege-in-Schleswig-Holstein/!5638793/ (accessed on 24 March 2023).
24. Baden-Württemberg.de. Verkehrsministerium Stellt Konzept Für Bahnsteighöhen Vor. Available online: https://www.baden-wuerttemberg.de/de/service/presse/pressemitteilung/pid/verkehrsministerium-stellt-konzept-fuer-bahnsteighoehen-vor/ (accessed on 10 July 2023).
25. Nießen, N.; Schindler, C.; Vallée, D. Assistierter, automatischer oder autonomer Betrieb–Potentiale für den Schienenverkehr. *ETR* **2017**, *4*, 32–37.
26. Salonen, A.O. Passenger's subjective traffic safety, in-vehicle security and emergency management in the driverless shuttle bus in Finland. *Transp. Policy* **2018**, *61*, 106–110. [CrossRef]
27. Cecon, F. Selbstfahrende Regionalbahnen und was Fahrgäste darüber danken. *Oöinitiativ* **2018**, *2*, 12–17.
28. Pagenkopf, J.; Schirmer, T.; Böhm, M.; Streuling, C.; Herwartz, S. Marktanalyse Alternativer Antriebe im Deutschen Schienenpersonennahverkeh. Berlin. 2020. Available online: https://www.now-gmbh.de/wp-content/uploads/2020/09/now_marktanalyse-schienenverkehr.pdf (accessed on 9 August 2023).
29. Deutsche Bahn, A.G. Infrastrukturzustands-Und-Entwicklungsbericht 2021. Available online: https://www.eba.bund.de/SharedDocs/Downloads/DE/Finanzierung/IZB/IZB_2021.pdf;jsessionid=469D5A56C212AF6275ED3F34ACE7F086.live21301?__blob=publicationFile&v=2 (accessed on 16 January 2023).
30. Ihme, J. *Schienenfahrzeugtechnik*; Springer Fachmedien Wiesbaden: Wiesbaden, Germany, 2016; ISBN 978-3-658-13540-9.
31. COMMISSION REGULATION (EU) No 1302/2014 of 18 November 2014 Concerning a Technical Specification for Interoperability Relating to the 'Rolling Stock—Locomotives and Passenger Rolling Stock' Subsystem of the Rail System in the European Union: TSI LOC&PAS. 2014. Available online: https://eur-lex.europa.eu/legal-content/EN/TXT/PDF/?uri=CELEX:32014R1302&from=DE (accessed on 10 July 2023).

vehicles

MDPI

Article

Securing the Future Railway System: Technology Forecast, Security Measures, and Research Demands

Simon Unger [1,*], Markus Heinrich [2], Dirk Scheuermann [3], Stefan Katzenbeisser [1,2,*], Max Schubert [2], Leon Hagemann [2] and Lukas Iffländer [4]

1 Chair of Computer Engineering, Faculty of Computer Science and Mathematics, University of Passau, 94032 Passau, Germany
2 INCYDE GmbH, 10117 Berlin, Germany; markus.heinrich@incyde.com (M.H.); max.schubert@incyde.com (M.S.); leon.hagemann@incyde.com (L.H.)
3 Fraunhofer SIT, 64295 Darmstadt, Germany; dirk.scheuermann@sit.fraunhofer.de
4 Deutsches Zentrum für Schienenverkehrsforschung beim Eisenbahn Bundesamt, 01219 Dresden, Germany; ifflaenderl@dzsf.bund.de
* Correspondence: simon.unger@uni-passau.de (S.U.); stefan.katzenbeisser@uni-passau.de (S.K.)

Abstract: The railway industry—traditionally a conservative industry with low adaption speed for innovation—is currently entering its digitization phase. The sector faces a challenge in integrating new technologies and approaches into the employed—often safety-critical—systems. Keeping the systems secure while conforming to the demanding safety norms creates previously unknown problems. In the last decades, the number of attacks on the railway system has increased. Furthermore, with standardized digital technologies, the attack surface will keep growing. Therefore, in this work, we look into the foreseeable future of the railway system and present 21 likely use cases. We analyze these use cases regarding possible threats, rate the severity of these threats, and deduce and rate necessary countermeasures. To this end, we model these use cases and the corresponding threats and countermeasures using Attack Graphs. We use a graphical solution for the risk and security analysis due to advantages over other methods, i.e., table-based solutions, like simplified presentation and an easier understanding of relationships, dependencies, and interactions between various elements. From these Attack Graphs, we extracted 14 commonly recurring attack strategies. After analyzing 49 countermeasures regarding their current maturity and further research and standardization demands, we identified 21 in need of further investigation. This implies that 21 necessary countermeasures to secure these future use cases require further research to apply to railway systems or require standardization. These results will help researchers focus on the necessary research and standardization and railway operators to ensure the security of their systems.

Keywords: railway system; Attack Graphs; technology forecast; security threats; security measures; standardization; research demand; transportation; critical infrastructure

1. Introduction

Future railway networks are expected to provide more capacity. For example, in Germany, the goal is to provide up to one-third more [1]. Since 2019, the German government has increased its annual investment in Germany's primary railway infrastructure operator, Deutsche Bahn's rail network, to realize this goal. Surpassing road infrastructure funding for the first time in 2022. The German government aims to double passenger numbers by 2030, primarily for climate policy reasons [2].

Significant performance improvements are necessary to achieve these objectives. This pertains to every aspect of the entire railway system, starting from traffic control and the utilization of the existing rail network to managing passenger flows at stations and aboard vehicles, facilitating a seamless connection to various transportation modes for the smooth and controlled movements of goods and passengers. Consequently, a comprehensive digitization of the technology and operational processes of the railway system is required. This objective not only fosters new technological trends but also enhances the capacity and attractiveness of the overall railway system. However, it also introduces new security vulnerabilities and risks, posing threats to the availability and operational safety of the railway system.

An analysis of physical attacks on the railway system by Iffländer et al. [3] shows that attacks have been drastically rising since the 1990s. Moreover, 35 out of the 127 attacks that have been analyzed have taken place in the last decade. This shows that the railway system is an increasingly popular target. Likely due to its high vulnerability and wide-spread and easily accessible network, it can be declared a so-called "soft target" [4].

In the "Forecast of security requirements and evaluation of possible security concepts for the railway system" project, we addressed the tension between technological trends and the challenges of threats by investigating the following research questions: (1) What technologies will be used in railway systems up until 2050? (2) What are the use cases for these technologies? (3) What threats are posed by these technologies and use cases? (4) How can we secure the systems against such threats? (5) Have the countermeasures to protect the systems against threats been researched enough and standardized for use in railway systems? To answer these questions, we began with a technology forecast up to 2050, describing use cases for a comprehensive overview of new technology. Using the technology forecast, we identified use cases covering all railway subfields, ranging from passenger and freight transport, over rolling stock and infrastructure, to train protection and signaling technology, as well as railway operations, traffic management, and maintenance. The use cases establish the connections between technologies, data flows, and benefits, directly utilized to identify security needs. To identify the threat landscape, we developed a methodology of Attack Graphs and derived an Attack Graph for every use case. After the identification of the threat landscape, we identified countermeasures necessary to secure the use cases. These countermeasures were then analyzed regarding their current technical states to identify those that required further research and standardization.

Using the technology forecast, we identified 21 use cases describing the used technologies and their relations. For each use case, we constructed Attack Graphs to depict the threat landscape, unearthing 14 recurring attack strategies that threaten these use cases. To secure future systems against those attacks, we identified 49 countermeasures; 28 of them are already established solutions ready for deployment in railway systems and are mandated by standards for implementation. That leaves 21 countermeasures grouped into 6 categories that are not sufficiently developed to be used in railway systems and/or are not mandated by standards for deployment.

Defining Attack Graphs as a graphical solution for the risk and security analysis mitigates the disadvantages of other risk and security analysis solutions, i.e., table-based solutions. The first advantage of Attack Graphs is that a visual representation of the security landscape enhances the understanding and communication of complex risk information and simplifies the identification of patterns, trends, and relationships among threats and risks. Furthermore, different risk scenarios can easily be explored by manipulating variables or parameters or adapting different countermeasures within the visual representation, enabling the assessment of the potential impacts of various risk factors and the evaluation of the effectiveness of different response strategies.

The remainder of the paper is structured as follows: After the introduction, Section 2 briefly introduces the related work to our goals. All derived use cases are described in Section 3. The developed methodology of Attack Graphs to support the risk analysis is summarized in Section 4. In Section 5, we identify frequently recurring attack strategies and countermeasures that still require research and standardization. Finally, Section 6 presents an outlook on the further work needed to evolve the derived countermeasures.

2. Related Work

To the best of our knowledge, no study exists that answers all of our research questions. However, there are studies that help to answer research questions 1 and 2; the most relevant are presented below.

It is necessary to know the threats to these future systems to identify the research and standardization demands of security measures for the future railway system. Furthermore, it is necessary to know the technologies used in future railway systems to identify the threats. As the railway system takes up innovation rather slowly, especially technologies that impact safety, some future technologies are already known, as they are being researched and developed. These technologies can be found in the literature and roadmaps of railway operators, federal offices, and research publications. These technologies include automatic train operation (ATO) [5,6], communication-based train control (CBTC), digital interlocking, the Future Railway Mobile Communication System (FRMCS), infrastructure modeling [7], the European Train Control System (ETCS) [8], train-to-train communication [9], predictive maintenance, global navigation satellite system (GNSS), innovative train integrity systems, fiber optic sensing, industrial robots, robotic assistants [10,11], and intelligent automation.

Furthermore, a report by Arup Rail titled "Future of Rail 2050" [12] describes how they envision the railway system up until 2050 and the technologies they deem necessary. Some of their envisioned technologies have security impacts, making them essential to our research. These include autonomous trains, real-time passenger information, intelligent robots, and drones. However, the study does not include specifics about technologies, and more notably, it omits insights on security issues, countermeasures, and the urgency for research and standardization. To further identify technologies and use cases, we looked into industries like the automotive industry.

To identify threats, multiple publications exist on specific technologies, which can further be adapted or envisioned for the railway industry. For example, Iffländer et al. covered physical attacks on the railway system that occurred in the past [3]. They analyzed physical attacks on railway systems worldwide and collected 127 events since the late 1800s. The attacks were structured and evaluated according to types of attackers, means of attack, targets of the attacks, and incurred damage. Their analysis indicates a significant rise in attacks and casualties since the 1990s. Furthermore, the means of attack are becoming more diverse. This study supports the security analysis regarding possible physical threats. However, it does not include digital attacks, security countermeasures, their state of research, or standardization.

3. Technology Forecast

The project team oversaw the development of the technology forecast. To expand the technology forecast predictions across the future railway system, the expansive expertise of the project partners in the railway and automotive sectors was extended by members from the areas of industry, research, and operational backgrounds to form a team of experts.

The determination of security requirements for the future railway system is based on a technology forecast that anticipates the digital technologies that will be utilized in the next 30 years as use cases.

Our technology forecast involved a three-stage process:

- Literature research and discussions within the project team.
- Workshops with an expert team stemming from railway operators, railway manufacturers, and academia.
- An exchange between the project and expert teams.

In stage 1, the existing specialist knowledge from the literature was documented, including the expectations of the project members. In stage 2, semi-structured workshops were held with the expert team, with the goal of

1. Finding features of the railway system that fall into the categories shown in Figure 1; and

2. Identifying future technologies that will be available until 2050 and could be integrated into the railway system.

For the technology forecast, 21 use cases (summarized in Section 3.3) were defined across 6 areas: passenger and freight traffic, train protection and signaling technology, rolling stock, trackside infrastructure, as well as railway operation, traffic management (dispatching), and maintenance; they describe the application of digital technologies in future scenarios (see Figure 1).

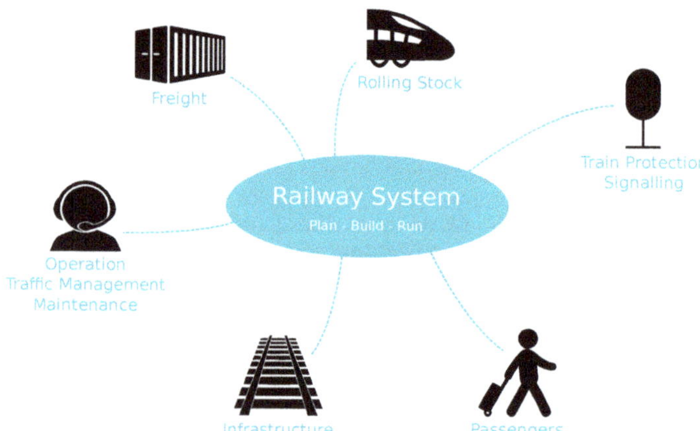

Figure 1. Areas of the railway system considered in the technology forecast.

While creating the technology forecast, two fundamental goals were pursued: sufficient completeness of the technology forecast and mutual validation.

3.1. Goal 1: Sufficient Completeness

Absolute completeness is challenging to achieve, with a forecast extending up to the year 2050. This particularly applies to the railway system that has numerous subsystems. The heterogeneity of the different subsystems caters to various technologies and use cases. The expert team members take interdisciplinary perspectives on the railway system and, thus, round off the technology forecast to obtain a complete picture. By consulting experts without a specific focus on the railway system, who are unaffected by the limits of the existing system, there is an opportunity to capture innovation that may remain hidden from internal experts. In addition, the technology forecast aims to assess the need for future IT security mechanisms, not to craft a perfectly accurate depiction of the railway or transport sector. Under these conditions, a certain amount of uncertainty in the technology forecast was deemed acceptable. Drawing knowledge from the literature and the expert team ensured sufficient completeness of our technology forecast.

3.2. Goal 2: Mutual Validation

The technological trends and expectations formulated by the expert team were utilized, among other purposes, for the validation of the technology forecast. To keep the mutual influence low, the expert team was divided into two groups to perform the workshops. Both groups independently mentioned the same or similar technologies and scenarios, thus validating each other and pointing to a reliable forecast. Accordingly, as expected, there were overlaps in the forecast technologies and use cases.

3.3. Use Cases

In this section, we briefly describe the 21 derived use cases. A detailed description of the derivation process and all the use cases can be found in Leining et al. (in German) [13]. The use cases we designed describe the employment of future technologies and their interactions. Therefore, we do not provide detailed architecture. Also, specific architectures can be different depending on the company or operator. However, for our research, this level of detail was not necessary. Understanding the technologies used and their communication medium is enough to identify the threats to the system. It is not necessary to know, e.g., how many routers are in the network. A single one might be enough for the system to be vulnerable, and multiple routers might only increase the attack surface.

Automated hump control. In marshalling yards, automated processes are used to assemble trains by shunting wagons. An automatic shunting locomotive pushes the coupled train over the hump, and the separation of wagons is achieved through the use of digital automatic coupling (DAC) on different destination tracks. To enable this, wagons must be equipped with digitally readable identifiers for detection along the track.

Automated planning and plan review. Planning processes today rely on subjective interpretations of rules written in prose, which can lead to inconsistencies. A complete and logical mapping of the planning rules using mathematical relations and logical connections enables the identification of the best and formally correct solutions while detecting conflicts with defined goals. The same automated process can be applied to plan reviews, using different logical connections to ensure independence and prevent logical shortcuts.

Cloud interlocking. The trend of centralizing interlocking systems continues towards centralized cloud-based solutions, offering infrastructure companies the ability to control their entire network from a single data center. This cloud-based interlocking system is a service, making it an attractive option for smaller networks reluctant to invest in their technology and expertise for a few interlocking systems. To ensure safety, it requires a safe platform and secure communication technologies between the data center and field elements. The advantages include centralized maintenance, backup resource availability, standard components, such as (commercial off-the-shelf (COTS)), and centralized personnel deployment.

Decentralized interlocking. Traditional interlocking systems for track safety are replaced with decentralized methods, where vehicles directly reserve routes or coordinate with each other through V2I or V2V communication. This eliminates the need for conventional interlocking systems. Accurate positioning of trains and infrastructure elements is ensured using technologies like GNSS, enabling secure localization and route planning stored on digital maps.

Emotion and intent recognition. Cameras are used to record the body language of individuals. Afterward, affective computing [14,15] can deduce emotions and intentions using facial expressions, postures, gestures, and, if available, language. This allows one to make conclusions about the passenger's mood or emotional state or it can be used to detect indicators of aggression or willingness to engage in violence early on, allowing for appropriate prevention and de-escalation measures to be implemented.

Remote-controlled rolling stock. Vehicles that can be controlled remotely from a central operation center depend on reliable wireless communication, such as FRMCS, along all possible routes to transmit extensive sensor data to the operations center and control data to the vehicle. Centralized remote control mitigates issues related to driver fatigue as driver relief can be handled more efficiently from a central location rather than dispersed along the route.

Freight route management. In the future, railway operators can offer a service encompassing end-to-end logistics support by integrating various transportation modes. This enables prioritization of low-emission transport routes and optimization of the route based on factors such as the required arrival time, temperature control, criticality for avoiding delays (e.g., just-in-time requirements), optimization of available capacities, and more. This requires a comprehensive digital representation of traffic and customer processes.

Predictive maintenance. Predictive maintenance aims to predict when a failure is likely to occur by analyzing the real-time operational data of the equipment. Predictive maintenance could be further enhanced to eliminate regular maintenance intervals and instead focus solely on addressing potential failures before they occur through timely maintenance interventions.

Intermodal freight handling. Intermodal freight handling describes the integration of different modes of transportation for efficient goods delivery. Beyond the virtual coupling of trains, individual wagons with their propulsion could be utilized for transporting standardized containers, which can operate on both road and rail or other track-bound systems, like the Hyperloop. The wagons are combined into larger units, utilizing platooning or virtual coupling to cover longer distances and optimize infrastructure utilization. Additionally, the possibility of transitioning to air transport is considered, and the control and coordination of these operations are explored through concepts like "driverless operations" or "cloud-based control".

Intermodal travel chain. Future mobility will be a service-oriented experience that combines multiple modes of transportation into a seamless journey with shared data and integrated payment systems. This intermodal approach allows travelers to book and pay for their entire trip simultaneously. Advanced technologies, such as biometrics, GNSSs, and short-range wireless communication, enable contactless ticketing, registering the entering and leaving actions of passengers, and automating luggage transfer between different transportation modes.

Contactless ticket control. The ability for passengers to validate their tickets on their smartphones, which is already available today, will be further developed to enable contactless authorization without any interaction with train staff. Upon boarding the train, the digital ticket will be recognized, employing technologies like Bluetooth, NFC, or satellite-based tracking. This process eliminates the need for manual ticket inspections and requires ticket data to be matched with a central authority through communication means like FRMCS or satellite communication.

Optimized travel and price arrangements. Travelers require access to digital platforms that provide comprehensive information on available travel options, pricing, and services, while also accommodating their specific preferences and offering real-time updates on disruptions and alternative routes. These platforms need to integrate data from various travel providers, connect with operational centers and vehicles, and leverage Big Data techniques to handle the complexity and volume of information. By accepting pre-defined travel profiles as input, the platforms can offer personalized recommendations and immediate alternatives to enhance the overall travel experience.

Optimization of passenger changeover time. Sensor-equipped trains monitor the occupancies of different sections and transmit the data to the next station. The data are then displayed to waiting passengers, allowing them to position themselves in less crowded areas before the train arrives. This reduces conflicts between boarding and exiting passengers while improving overall comfort by maximizing available space and minimizing travel distances within the train.

Identification of persons. Identifying individuals has multiple benefits, including analyzing their behavior and providing personalized offers. It also aids in tracking individuals involved in criminal activities or those who are helpless or missing, allowing for targeted responses and the involvement of relevant authorities.

Travel and life management. In the future, railway operators can offer a service for end-to-end customer guidance through a connected multi-modal transportation system. This allows for a preference for public transportation while optimizing schedules and capacities to reduce the drawbacks of rigid timetables. Last-mile options like car sharing, autonomous taxis, and bicycles complement the offering. The foundation is a comprehensive digital representation of traffic and customers, enabling the calculation of an optimal route based on individual requirements, habits, and preferences.

Passenger guidance through the train station. Passengers on the platform should receive real-time information on their mobile devices to minimize infrastructure and provide convenience. Location-based services enable tailored information specific to the platform, utilizing the Internet of Things (IoT), mobile networks, wireless technologies (e.g., LoRa®)

for longer distances, and shorter-range options (e.g., Bluetooth or ZigBee) for local communication. This approach improves accessibility and ensures passengers remain informed without relying on physical display boards.

Steering action for scheduling. The control center's dispatch utilizes standardized interfaces for immediate operational interventions and relies on a digital twin of the railway system to generate precise forecasts. High computing power from technologies like quantum computers or exascale computing enables real-time optimization and the ability to respond dynamically to transportation demands. The proposed measures are promptly communicated as speed profiles and training paths to vehicles, interlocking, and Radio Block Centres (RBCs) for rapid implementation.

Virtual Coupling. For freight and passenger transportation, transportation units often use shared methods before eventually parting in different directions, or different transportation units from different methods come together. The process of physical coupling and decoupling is time-consuming. The concept of virtual coupling involves avoiding mechanical coupling processes. Instead, automatic steering directs the transport units along predefined routes (together); they are coupled and decoupled to and from different ways at predefined positions.

Fully automated driving. In the future, train journeys will be fully automated, with vehicles performing the operations. Communication with the track infrastructure is essential to avoid conflicts in occupancy and prevent accidents. The vehicle automatically travels along the track to its destination, considering vehicle and track limitations within a speed profile.

On-site information. Upon arrival at the track, passengers need to be supplied with current information, e.g., regarding a sequence of wagons, delays, or track changes. Passengers can obtain information using their mobile devices and contact service applications to navigate to the next sign panel.

ETCS Train Protection. From today's perspective, implementing already standardized ETCS L3, Hybrid L3, and L2oS will significantly influence the train's control system, moving away from traditional track vacancy detection systems. The positioning and train integrity determination are performed continuously on the vehicle side and transmitted to the interlocking or RBC. Train spacing will transition from the fixed block to the moving block, with movement authorizations relayed from the interlocking/RBC to the vehicles.

4. Attack Graphs

In the next step, we conduct a security risk analysis of the developed use cases. To support the risk analysis, we developed the methodology of Attack Graphs and built a software tool to automate the calculations. The underlying methodology and the software tool have been described in more detail by Unger et al. [16] and Heinrich and Iffländer [17]. In principle, the tool allows adapting attributes freely to the current practice of the domain under consideration, existing standards, and the risk affinity of the analyzing organization. The German pre-standard DIN VDE V 0831-104 [18] uses the attributes "resources" and "knowledge" known from the IEC 62443 [19] series and replaces motivation with three railway-specific risk factors, of which, we use "location" for our analysis. This leads to the attribute vectors (resources, knowledge, location).

Here, we only introduce the fundamental concepts and focus on the choices made within the Attack Graph methodology to apply them to our analysis. The tool reduces the amount of work and the susceptibility to errors, increases the traceability and meaningfulness of the analysis, and is freely available as open-source software (https://github.com/INCYDE-GmbH/drawio-plugin-attackgraphs Accessed on 22 September 2023).

An Attack Graph specifically analyzes one use case (see Section 3.3) at a sufficiently specific level of abstraction. For our study, we created 21 Attack Graphs accordingly. One graph depicts the analyzed asset by a trapezoid (see Figure 2). Attacks on the asset lead to various consequences or damage events, which are represented in the graph by rectangles with rounded corners. We consider the following consequences in our analysis:

- Financial damage to the operator, an individual, or a customer;
- Reputation damage to the operator, manufacturer, or an individual;
- Violation of laws and regulations;
- Violation of privacy;
- Infringement of property;
- Restriction of a critical service;
- Provoke chaos;
- Market manipulation;
- Obstruction of justice;
- Violation of physical integrity.

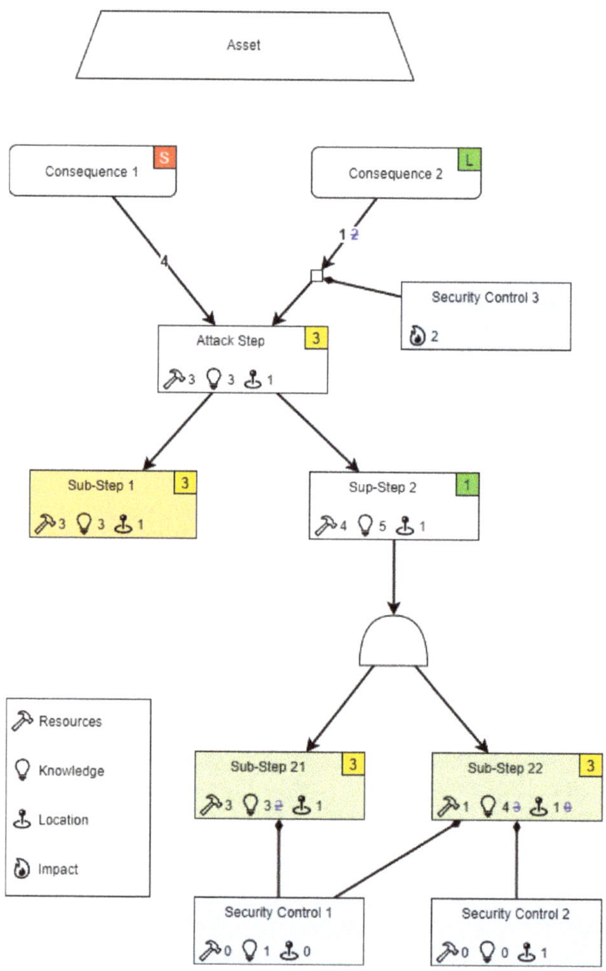

Figure 2. Example of an Attack Graph.

4.1. *Refinement of Attacks*

Damage events are further refined by considering particular attacks. Each damage event can be triggered by one or more attacks ("attack step" in Figure 2). The graph shows this relationship via a directed edge from the consequence to the attack step. By visualizing this link, the Attack Graphs increase the traceability of the risk analysis since they can graphically represent possible n:m relationships between attacks and consequences. The Attack Graphs support the analysis by graphically breaking down the attacks into sub-steps ("Sub-step 1" and "Sub-step 2" in Figure 2) to refine the analysis and the subsequent risk assessment. Sub-steps may be combined into attacks using logical disjunctions (or) and conjunctions (and). The methodology allows for nested logical operators and is open for extension with other logical operators (e.g., XOR). A directed edge between the nodes represents a division of a step into the sub-steps. Corresponding nodes between two attack steps represent disjunctions and conjunctions (e.g., the AND gate in Figure 2). A direct connection between two attack steps implicitly represents a disjunction. The breakdown of the attack continues iteratively until reaching attack steps whose probability of occurrence (resources, knowledge, and location) can be estimated with sufficient precision.

4.2. Risk Assessment

After setting up the Attack Graphs, the risk assessment process follows a bottom-up approach, starting with the leaf nodes and aggregating results upward to compute the final assessment. In this analysis, the tool evaluates the leaf nodes of the Attack Graphs by a selected vector of attributes to model the probability of occurrence. As described before, we use the attribute vectors (resources, knowledge, location) taken from the German pre-standard DIN VDE V 0831-104 [18], which is based on the IEC 62443 [19] series. The tool visualizes the attributes with the help of icons (see Figure 2). The added value of Attack Graphs comes from atomic leaf nodes, as their attribute assessments are more straightforward than more complex compound attacks. However, this means that it is necessary to aggregate the leaf node evaluations along the path to the damage event to reassemble the sub-steps into an overall assessment. The software tool used for creating Attack Graphs supported analysts through the prepared and automated linking of the sub-steps according to the logical operators defined in the breakdown. A detailed description of the required functions is presented by Unger et al. [16]. A disjunction of attack steps assumes that the attacker chooses the most easily executable step out of several possible sub-steps to implement an attack step. A mathematical order relation over the attribute vectors allows comparing the feasibility of the partial steps and determining the highest feasibility as a scalar from the attribute vector (R, K, L), as depicted in the top right corner of each attack step. For example, in Figure 2, the evaluation of "Sub-step 1" carries over to the "attack step" because it has higher feasibility compared to "Sub-step 2". The combination of several sub-steps leads to the more complex (and lower) feasibility of the resulting step. The resources and knowledge required for a successful attack have increased, as shown in Figure 2. For illustration, the interposed "AND" node adds the attributes of the child node "Sub-step 21". The software tool aggregates the attribute vectors automatically toward the parent attack step (against the direction of the edges). Experts also evaluate the impact of an attack or its influence on a specific consequence at the transition from the attack step to the consequence. The estimated impact is represented as the edge weight of the edge between the attack step and the consequence. An aggregation function in the consequence node accepts each child node's attribute vector and the impact (edge weight). It uses a risk matrix to determine the attack with the highest risk for the damage described in the node (the letter in the top right corner of consequences). The values shown in the figures are for illustration purposes and do not reflect a specific use case.

4.3. Countermeasures

The risk analysis and assessment of the threats can consider countermeasures, which also reduce the risk. Here, we assume an n:m relationship between the attack and countermeasure since, typically, a single countermeasure can protect against several attacks and, conversely, multiple countermeasures can protect against the same attack, or only the combination of several measures can provide adequate protection. Figure 2 shows the effects of two countermeasures on two attack steps and their impacts. The countermeasure evaluation uses the same attribute vectors as attack steps to determine the levels of mitigation they provide to an attack or impact. The tool displays the original assessment in blue and crossed-out font, while the values printed in black account for the influence of the countermeasures and, thus, reflect the risk reduced by the measure. Both "Security Control 1" and "Security Control 2" reduce the attack feasibility of "Attack Step 1" with the added effect in the example, which eventually leads to a risk reduction. On the other hand, only "Security Control 2" acts on "Attack Step 2", so there is a minor reduction in feasibility with the same initial evaluation as "Attack Step 1". "Security Control 3" mitigates the impact of "Attack Step" on "Consequence 2".

4.4. Outcome

All 21 analyzed Attack Graphs, including the attack paths, the attribute assessment, and the countermeasures, are available online (https://github.com/INCYDE-GmbH/attackgraphs accessed on 22 September 2023). In workshops, security experts evaluated attributes for each attack step.

The greater goal of our analysis was to identify future standardization and research demands, i.e., gaps in current technology and requirements in the protection against the anticipated threats that we describe in Section 5.1. We strived for completeness in identifying the gaps in the countermeasures. Still, we did not provide a complete enumeration of measures to reduce the risk of each threat below an acceptable threshold ("low", in our case). The latter would require demonstrating the completeness of the application of existing countermeasures, which is not necessary to achieve the goal of our analysis.

5. Results and Discussion

Based on the use cases described in Section 3.3 and the resulting Attack Graphs, we identified threats to the future railway system. Analyzing the threats, we discovered that all threats correspond to 14 different categories described in Section 5.1. In order to mitigate these threats, we further identified countermeasures, particularly those that induced research and standardization demands (see Section 5.2).

5.1. Upcoming Threats to the Future Railway System, Inducing Research Demands

After analyzing the 21 generated Attack Graphs, including the identified threats, we grouped them into the following 14 categories.

Unauthorized data access. Unauthorized access to data storage or other components in the railway system poses a threat to large and small datasets, such as configuration or control data, which are crucial for automated railway operations. Manipulating data through unauthorized access to system components and interfaces can entail significant risks, even if only a few small datasets are affected.

Interception of data. Even without manipulation, the interception of confidential data poses a significant threat that can carry high risks. In the context of railways, this primarily concerns passenger data, such as travel profiles or other personally identifiable information (PII). Unauthorized interception of such data during the use of services, like passenger guidance or travel management, represents a violation of the passenger's privacy. An analysis of the use cases indicates that as there is a rise in individualized, profile-based assistance for passengers coupled with data protection regulations, new risks arise that require adequate concepts to protect privacy (ideally "Privacy by Design").

Tampering with Wireless Transmission. In the future, communications in railway systems will increasingly rely on wireless technologies, such as mobile networks, satellite communication, or short-range technologies, like Bluetooth or Sigfox. Wireless communication plays a significant role, especially in machine-to-machine (M2M) communication and, therefore, it needs to be protected. However, remote-controlled vehicles are also attractive targets that will be used, for example, for shunting operations or in ATO systems. Attack scenarios include disrupting wireless communication, such as blocking M2M transmissions or preventing remote control. Furthermore, false data packets can be injected, leading to incorrect occupancy or the release of tracks in M2M communication, or in the case of remote control, taking control of the vehicles.

Attacks by quantum computers. Quantum computers utilize qubits instead of classical bits, allowing for coherent superposition of the 0 and 1 states. This provides an advantage in computational speed over classical computers, enabling exponentially complex calculations to be performed much faster. However, quantum computers do not pose a significant threat at present because they still have very few qubits. Many techniques for ensuring integrity and authenticity rely on calculations that become exponentially more difficult as the bit length increases. Nevertheless, research institutions and industries are continuously working toward expanding and improving their quantum technologies, which may pose a future threat to security. Therefore, it is essential to use methods that ensure integrity and authenticity, which are robust enough to counter potential threats from quantum computers.

Manipulation of (large) databases. As railway operators seek to improve their systems, expand their offerings, and provide personalized services, data collection becomes increasingly essential. The data are necessary for various applications, such as enabling predictive maintenance, training artificial intelligence (AI) models, and offering tailored products to individual customers. Therefore, the storage of large datasets plays a crucial role. It is vital to ensure the data are neither lost nor compromised in integrity. Failure to renew system components appropriately by predictive maintenance can lead to financial or, in the worst case, personal harm. Manipulating training data for AI systems can also compromise the intended function of the AI. Customer-related data pose significant risks, especially in future use cases like intermodal travel, where sharing data with other transportation service providers may be necessary to a certain extent. However, even in more conventional applications, such as contactless ticket inspection and passenger identification, it is crucial to ensure the privacy of individuals. Paradoxically, the provision of anonymized data can still threaten individual privacy since, with an increasing amount of available data, it is possible to identify individuals through data correlation, even if all the data are anonymized. Therefore, it is essential to ensure the confidentiality, integrity, and authenticity of datasets.

Attacks on the digital capture of the environment. Like many other industries, the railway sector relies on an ever-increasing amount of digital data of various types. Even without attackers, creating and maintaining an accurate digital representation of the analog world is already a challenge. This challenge is further amplified by the use of digital twins, which monitor the digitally cloned system in near real-time for changes (such as the movement of trains or people and reactions to environmental conditions). Through cyberattacks, the dimensions of deliberate and concealed manipulations of the real-time digital representations of the analog world come into play. Attackers could manipulate infrastructure data in digital maps (used, e.g., by ATO) to provoke accidents or disrupt railway operations on a large scale. For example, suppose measurement data from autonomous trains are manipulated and do not accurately reflect reality. In that case, they can impact functional safety and potentially enable extortion of the operator. Applications such as automated or autonomous driving heavily rely on the accurate perception of the environment. A future challenge will be to ensure high trust in the correct collection and retention of different datasets. Technical approaches to address this challenge may include validation through historical data points or integrating different measurements.

Manipulation of machine-to-machine communication. As the railway system undergoes digitization, the technology forecast also predicts the automation of various processes. This leads to an increase in communication between system components, such as vehicle-to-vehicle (V2V), vehicle-to-infrastructure (V2I), and other forms of M2M communication. It opens up new targets for attackers since actions previously carried out mechanically can now be triggered digitally. One potential risk exists in opening a DAC to stimulate train separation, which could, in the worst case, pose a safety hazard to following trains. Regarding decentralized interlocking and driverless operations, M2M communication plays a significant role. It can be abused to disrupt or transmit false track occupancy between infrastructure, vehicles, and control centers. Moreover, components not originally part of the system (the attacker's devices) can be introduced to execute these attacks. Therefore, this communication must be protected against manipulation by ensuring the authenticity of communication partners and the integrity of messages.

Attacks on Machine Learning. Predictive maintenance is the most prominent example of using machine learning (ML) in the identified scenarios. ML and AI are also employed in other scenarios to fulfill partially critical functions. A threat arises from the fact that decisions are increasingly entrusted to AI, but such decisions are not equally as transparent and traceable as those made by humans. Attackers can exploit this situation to manipulate decisions in their favor. Relying entirely on predictive maintenance, e.g., by eliminating regular maintenance intervals for points, can lead to point failures. Obstacle detection based on the AI-based analysis of camera images can be deceived by existing obstacles to provoke collision. Such attacks can be carried out by manipulating the AI model if its integrity is not adequately protected.

Adversarial machine learning [20] has shown that ML can be deceived in classification through cleverly manipulated inputs. In such scenarios, the attacker leverages knowledge about the algorithm and model to orchestrate an attack. In image recognition, in particular, the manipulations are so subtle that a human observer cannot detect them.

Physical attacks. Currently, providing physical protection against cyberattacks on railway system components is particularly challenging because rail transportation inherently has openly accessible critical components. These include field elements of signaling technology and vehicles with passenger Wi-Fi. The physical attack surface will further expand as monitoring elements become more centralized. For example, the automation of shunting and train operations, as well as remote-controlled train driving, reduce the presence of the train driver, making it more challenging to detect physical manipulation. Signaling technology, in particular, is characterized by a physical separation between functional safety and IT security, which complicates the implementation of a comprehensive defense strategy.

Cyber extortion. Currently, infestation with ransomware and the associated demand for ransom and protection money are dominant patterns of cyberattacks. The analysis of the Attack Graphs indicates that there will be an increased vulnerability of railway system operators in the future, where besides ransomware, other extortion scenarios may also be possible, such as extortion for protection money under the threat of otherwise conducted cyberattacks. This affects both the operational components (signaling technology, dispatching) and the customer relationship (passenger and freight transport). Adequate IT security concepts must be implemented to counter all cyberattacks associated with extortion in the future. The vulnerability to cyberattacks and the threat of cyberattacks result from the increasing automation of processes and the use of digital datasets concerning the infrastructure and the customers. The more crucial a process or dataset is for the availability of critical rail transport services, the more strongly the cyber extortion affects it, as the pressure on the operator is heightened.

Malware. Introducing malicious software into the railway system can cause significant damage in various ways. The use of ransomware followed by extortion is just one example. Other types of malware can directly cause malfunctioning components in the railway system, posing high risks of damage. This includes generating false information about the system's state or manipulating users into engaging in harmful interactions.

Manipulation of safe localization. A variety of processes within the railway system already benefit from location information. These include, for example, the self-localization of customers using the operator's travel app or initial applications in railway operations through driving recommendations transmitted to the train driver on a separate mobile device. In the future, additional services are expected to be introduced that will facilitate dispatching in both passenger and freight transport and provide customers with real-time tracking of their goods. Additionally, safety-related applications such as train positioning and train integrity are intended to be added to reduce the equipment required for the infrastructure. All these applications rely on a purely radio-based positioning system, collectively referred to as GNSS. However, all radio-based services, including GNSS, have the shared vulnerability of being susceptible to jamming. When jamming occurs, the location information becomes unavailable. Processes entirely reliant on GNSS positioning are no longer operational during such incidents. Jamming attacks on GNSS can be relatively quickly executed due to the relatively weak signals from satellites. A more challenging and critical security concern is the spoofing of location information. Spoofing involves falsifying the location information reporting a different location than the actual position, such as a different track or section. In safety-critical applications, this can have catastrophic consequences. Current railway systems, such as locomotives, lack sufficient measures against jamming and spoofing. With the future intensive use of localization methods on vehicles, it is essential to employ methods for protection, detection, and response to jamming or spoofing of localization. Therefore, a resilient architecture implementing these functions is crucial for maintaining safe and available railway operations. The examination of these measures and their expansion to less critical applications must follow to ensure the availability and resilience of convenience-enhancing services against attacks.

Denial-of-service attacks. Jamming, described next, belongs to the class of denial-of-service (DoS) attacks, where the availability of components in the railway system is disrupted. These attack scenarios are not limited to the disruption of radio signals alone. Numerous other attack methods can achieve a DOS. Examples include flooding servers with service requests or mechanically disrupting data transmission by damaging transmission paths.

Jamming. The technology forecast predicts a continued increase in wireless communication systems used in various applications. Functions such as train operation, safety, passenger information, freight transport, and intermodality rely on wireless communication. These functions are realized through different ranges, bandwidths, and network coverage technologies, including 5G networks, FRMCS, satellite communication, Bluetooth, near-field communication (NFC), and others. However, all wireless transmission technologies are vulnerable to jamming attacks since they utilize radio waves that are accessible to everyone. Therefore, jamming will remain an attractive attack scenario in the future because current defense strategies and resilience concepts do not sufficiently hinder or prevent successful jamming attacks.

5.2. Countermeasures with a Further Need for Research or Standardization

Our work aims to investigate necessary countermeasures that keep the future railway system secure and identify those countermeasures where further standardization and research are required. Hence, after creating the 21 Attack Graphs, we identified countermeasures that reduced the risk to an acceptable amount. Countermeasures were given values of how much they reduced the risk by rating how much they either increased the necessary knowledge or resources to perform the attack successfully or how they decreased the impact caused by a successful attack. This resulted in a list of 49 desired countermeasures and direct visualization of resulting risk mitigation in the case of successful applications to the identified abuse cases.

These 49 countermeasures mitigated at least one threat described in Section 5.1. Some of these relations resulted from a single attack path in one of the 21 Attack Graphs. To discuss all relations in detail, we would therefore need to include the Attack Graphs, which we restrain from, as it would exceed the scope of this article. However, the Attack Graphs that describe the threats are already published and available online (https://github.com/INCYDE-GmbH/attackgraphs accessed on 22 September 2023).

Subsequently, we estimated the maturity of the 49 countermeasures, i.e., to which degree the countermeasure is already in use or ready for use in current railway systems. We use three maturity levels based on the well-known technology readiness level (TRL) [21]:

A. The technology is in an early development or research state, meaning that there is, at most, a laboratory experiment (TRL 1–4).
B. First prototypes of the technology exist (TRL 5–7).
C. The technology is well-known and used; there is no need for further standardization or research (TRL 8–9).

To estimate the maturity level, we use three different standards to verify if they also demand the respective countermeasures. The standards used are as follows:

1. The IEC 62443 series [19];
2. The BSI IT-Grundschutz [22]; and
3. The concretization of requirements for critical infrastructures (CIs) [23].

After determining the maturity level, we discovered that 21 out of 49 countermeasures have level "A" or "B" and, thus, are not yet mature enough to secure real-world railway systems. Thus, for these 21 countermeasures, there is either some research demand to develop the technology further, they need to be better reflected in standards, or both. We further group the 21 countermeasures into six categories in Figure 3:

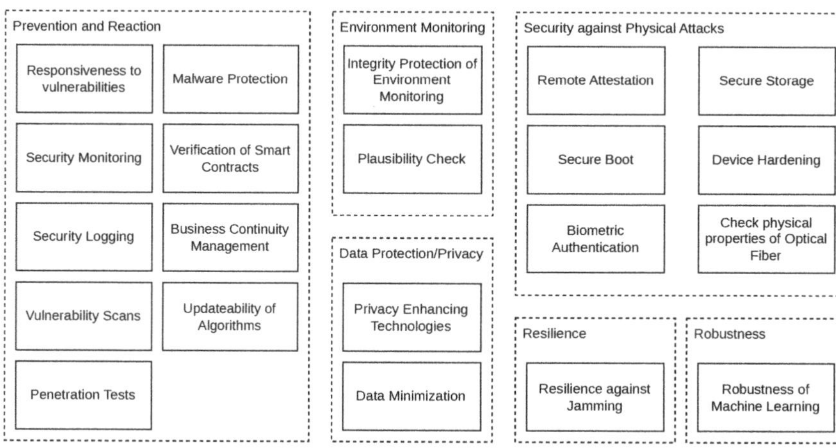

Figure 3. Countermeasures with a further need for research and standardization accumulated into six groups.

A total of 28 countermeasures that are already in use or deployed and are required by the standards have been identified. For some use cases, these existing countermeasures are sufficient, according to our risk evaluation, to reduce the risk to an acceptable level. In this case, the search for possible countermeasures was stopped, with an acknowledgment that there might be more or other well-established technologies to reduce the risk. However, as the goal was to identify the necessary demands not covered by state of the art solutions, we did not aim for the completeness of existing and standardized technologies.

Our work's purpose is to identify the necessary research and standardization needs, rather than determining the potential for improvement. As such, we did not conduct a security evaluation for countermeasures that are currently in use or anticipated to be used. We operate under the assumption that countermeasures are correctly and adequately applied to mitigate risks. This holds true for both existing and future countermeasures. For example, we assume if encryption is necessary to reduce the risk of a threat; an encryption algorithm is chosen that fulfills the system requirements, including the security requirements, instead of an outdated or even a broken one.

In the subsequent sections, the groups are described, including the necessary actions to raise the maturity level of the respective countermeasures.

5.2.1. Prevention and Responsiveness

A large group of countermeasures is devoted to activities that prevent cyberattacks and the ability to react to newly disclosed vulnerabilities and a changing attack surface. Concepts for the ability to react to detected security issues already provide vital countermeasures that help to prevent further successful cyberattacks.

An essential class of cybersecurity threats is provided by malware infection. Therefore, malware detection and prevention (see, e.g., [24]) play important roles. While most standards already demand malware protection, an adequate implementation cannot yet be found in many systems of railway applications due to long system life cycles (legacy systems), lack of update possibilities, the requirement for readmission in the case of software changes, and the absence of software solutions for certain types of platforms. Research should investigate and define appropriate levels of protection against malware, which does not necessarily result in the development or adaption of anti-virus software.

If an attack is successful, systems should be fail-safe or fail-operational. This property is supported by business continuity management (BCM), which was already considered for IoT [25]. As demanded by BCM (compare ISO 22301 [26]), disaster recovery plans are not yet fully developed and tested for railway applications. Relevant systems need to catch up to state of the art BCM to maintain the availability of railway transportation, especially the parts that are considered CIs'.

Critical systems must be frequently examined for potential vulnerabilities that are helpful for an attacker. Vulnerability scans provide this. Furthermore, prevention can be reached by testing the execution of cyberattacks. In the past, these penetration tests were also considered in the area of IoT networks [27]. Both vulnerability scans and penetration tests are ready for deployment. However, field tests in railway applications are still pending to elevate the maturity level.

Further preventive countermeasures are provided by the continuous observation of the actual system's security state. In the first step, security logging is performed to collect and note all possible security-relevant system state information. Security monitoring is not yet applied in railway systems, even though there are proposals for doing so [28]. This provides functionalities like intrusion and anomaly detection as well as intrusion prevention. Security monitoring was already investigated along with networked control systems [29].

New developments may require the update of used algorithms, e.g., in the context of cryptography. Therefore, the ability to update algorithms is another countermeasure in the responsiveness category. As system life cycles can last over decades in railway transportation, it is necessary to provide a sufficient margin in the system's capabilities to incorporate future security controls, such as improved cryptographic algorithms.

To detect and prevent malicious modifications of smart contracts, the verification of smart contracts is another important preventive countermeasure. Formal methods to perform this task have already been investigated [30].

Most of these countermeasures can already be found in several standards. However, others are not standardized sufficiently, like the ability to update algorithms, or are not even mentioned at all, like the verification of smart contracts. Hence, to increase the maturity, some countermeasures still need to be standardized, while others need to be adapted for and applied throughout the railway systems.

5.2.2. Environment Monitoring

Proper and secure monitoring of the environment provides important security protection in railway systems. This includes the provision of integrity for any measured environmental data and conducting plausibility checks, i.e., checks for abnormal data values that are suspicious of manipulation.

Regarding the integrity of environmental data, some rules can already be found in all three investigated standards. However, some security issues were identified that were not adequately addressed by the existing standards, leaving them exposed to significant risks.

Furthermore, the technologies and methods used to ensure the integrity of environmental data are not applied consequently enough throughout the railway industry. In particular, in scenarios where safety-critical decisions are (semi-)automatically drawn from digitally captured environmental data, harmful consequences need to be avoided by a proper security concept. Examples include track clearance detection of automated trains, a train's digital twin used for predictive maintenance (implying safety responsibility), or detecting people (not necessarily identifying them) to avoid congestion in crowded areas.

No specific or relevant solution or regulation can be found to validate data using multiple input sources. This is because the technology for environmental recognition is still being developed and tested; it is not widely deployed. Hence, technologies and methods for plausibility checks must be further researched and standardized. This implies that these technologies must also be applied in the railway domain.

5.2.3. Security against Physical Attacks

Physical security covers all countermeasures, ensuring proper and secure functionality of technical processes or mechanisms protecting the system against physical access by unauthorized persons.

A proper system start may be provided, e.g., by Secure Boot, to ensure that attackers with physical access to a device cannot change the executed programs to mount an attack. While Secure Boot may be available for many systems, it must still be applied to many of our use cases, especially in operational technology (OT) and consumer devices. For protection against unauthorized access, remote attestation [31] can be used to verify to third parties that the software running on a particular device remains unaltered.

A further countermeasure is provided by biometric user authentication, in particular, voice recognition in connection with speech commands, together with ML. Our identified standards only briefly touch on the topic of biometrics but without closer consideration of our specific context or specific methods. Voice recognition systems for different application areas are already in use, but solutions to our specific context of ML still need to be developed.

By hardening devices or providing secure storage for cryptographic keys, further unauthorized physical access can be prevented. Trusted platform modules (TPMs) are standardized solutions for secure key storage. However, other standardized solutions must be provided since TPMs are not always feasible.

Fiber optic sensing (FOS) is an emerging technology in the railway sector [32]. While more decisions are based on the measurement results of FOS, the attack surface on the system increases. Physical manipulation of FOS could be countered by continuously monitoring the (physical) properties of the fiber up to the extent of a physically unclonable function (PUF) [33,34]. This group of countermeasures also includes particular technical disciplines away from current cybersecurity methods, which are out of scope for our considered standards. Before using these countermeasures, both the physical properties to be inspected and the inspection methods require more standardized guidelines.

Considering physical attacks and the introduction of countermeasures for physical security that already started during the engineering process, the security functions must run on the same hardware as the actual service [35] to provide sufficient physical protection. This can be challenging, especially in OT domains such as railway transportation, which is subject to the authorities' control, which can be challenging. While the remaining risks stemming from physical attacks are widely accepted today, this assessment will likely change due to the increased scalability of digital–physical attacks. Foundations to encounter such attacks are subject to research [28,35,36], but solutions have not yet transpired into operational systems. Hence, more research and development are needed to implement adequate protection against physical attacks on digital systems.

5.2.4. Data Protection and Privacy

As part of the transportation system, the railway industry handles the sensitive data of individuals and companies alike, including personal data, payment data, and location data of individuals. The data need to be protected to ensure the rights and privacy of all involved, while business models of stakeholders depend on gathering and sharing such data.

Privacy-enhancing technologies (PETs) aim to protect sensitive data while maintaining usability for interested parties [37–39]. A further strategy is data minimization, i.e., the avoidance of using unnecessary data. Privacy in connection with public transportation was already covered in the literature [40,41].

Neither PETs nor data minimization is covered in the standards relevant to the railway domain. Data minimization is only covered in the BSI IT-Grundschutz-Compendium [22] in regard to web browsers. However, the data mentioned above are not only handled via web browsers; possible scenarios include more data of those kinds from various sources and for mixed-use. Parts of the data might also have to be shared with other companies. Apart from the necessary standardization that needs to be conducted, methods and technologies need to be further improved to handle private data accordingly.

5.2.5. Resilience against Jamming

Wireless communication plays a significant role in the railroad area, and this communication is heavily susceptible to jamming attacks. Therefore, resilience against jamming is a vital countermeasure.

For CIs, detecting jamming attacks already provides a vital step [42]. Fallback strategies—either to other carriers or backup processes, to maintain the railway service's availability—need to be developed and implemented to mitigate the consequences of a jamming attack. Beamforming seems to be a promising way to realize resilience against jamming [43–45].

However, resilience has not yet been covered in the relevant standards. The possibility of counter-jamming must be researched, developed, deployed, and standardized to increase maturity.

5.2.6. Robustness of Machine Learning

For a couple of use cases in the railroad sector, ML will play an important role. Hence, the robustness of ML is another essential countermeasure. Defenses against attacks on ML are often referred to as adversarial machine learning [20,46]. ML is closely connected with AI; a recent report detailing the current research requirements in this area was published by ENISA [47].

Adversarial ML and resilient AI are currently subjects of investigation. To ensure the correctness of future ML applications, further research is necessary in the railway industry, like using ML for predictive maintenance.

6. Conclusions and Future Work

Our research aimed to identify the research and standardization demands of security measures for future railway systems, as no such study exists today. Therefore, along with an expert team stemming from railway operators, railway manufacturers, and academia, we envisioned the potential landscape of railway systems up to 2050. We used 21 use cases by utilizing studies like the "Future of Rail 2050" by Arup Rail [12]. The use cases describe the utilized technologies and their connections, data flows, and benefits. These descriptions formed the basis of our subsequent security analysis, as no security analysis for these use cases has been conducted before. Therefore, we created an Attack Graph for every use case describing possible attack vectors and impacts. An analysis of these Attack Graphs showed that there are 14 frequently recurring attack scenarios. To ensure the security of future systems, we identified 49 security measures that are necessary to mitigate the threats and reduce the impacts of the attacks to an acceptable level. Another contribution of our work involved identifying necessary research and standardization demands by rating the maturity of countermeasures. This resulted in 21 countermeasures, which we grouped into 6 categories based on the urgency of action required.

To adapt to upcoming security vulnerabilities, the prevention and detection capabilities of the systems need to be improved by security monitoring, BCM, and increased flexibility. Systems that rely on the perceptions of their environments must be hardened against falsification to maintain safe operation; this is accompanied by the need for increased protection against physical attacks, including measures like secure storage, Secure Boot, and remote attestation. The protection of PII is not yet in full force and it needs to be improved. Attention is needed on the resilience of all utilized wireless transmission technologies against jamming attacks, as well as the robustness of employed ML and AI against manipulation, including adversarial ML. Action must be taken by manufacturers, operators, and authorities of railway transportation authorities alike.

To conclude this article, we demonstrated the pressing need for research and standardization to secure the railway system of tomorrow. Future works should focus on investigating the identified security measures with the need for action, adapting them to railway systems. Furthermore, once the countermeasures are developed, they need to be standardized to ensure the security of all railway systems.

Author Contributions: Conceptualization, L.I., M.S. and S.K.; methodology, D.S., M.H., M.S., S.U. and S.K.; validation, D.S., L.H., M.H., M.S., S.U. and S.K.; investigation, D.S., L.H., M.H., M.S., S.U. and S.K.; resources, M.H. and S.U.; writing—original draft preparation, D.S., M.H. and S.U.; writing—review and editing, L.H., L.I., M.S. and S.K.; visualization, D.S., M.H. and S.U.; supervision, D.S., L.H., L.I., M.H., M.S., S.U. and S.K.; project administration, L.I., M.H. and S.K.; funding acquisition, M.S. and S.K.; All authors have read and agreed to the published version of the manuscript.

Funding: The presented work is part of the "Security Requirements Forecast and Evaluation of Possible Security Concepts" project, commissioned and financed by the German Centre for Rail Traffic Research at the Federal Railway Authority (2020-22-S-1202).

Data Availability Statement: Publicly available datasets were analyzed in this study. The data can be found here: https://github.com/INCYDE-GmbH/attackgraphs. accessed on 22 September 2023.

Conflicts of Interest: The authors declare no conflict of interest.

Abbreviations

The following abbreviations are used in this manuscript:

AI	artificial intelligence
ATO	automatic train operation
BCM	business continuity management
CBTC	communication-based train control
CI	critical infrastructure
COTS	commercial off-the-shelf
DAC	digital automatic coupling
DoS	denial-of-service
ETCS	European Train Control System
FOS	fiber optic sensing
FRMCS	Future Railway Mobile Communication System
GDPR	General Data Protection Regulation
GNSS	Global Navigation Satellite System
GPS	global positioning system
IoT	Internet of Things
M2M	machine-to-machine
ML	machine learning

NFC	near-field communication
OT	operational technology
PETs	privacy-enhancing technologies
PII	personally identifiable information
PUF	physically unclonable function
RBC	Radio Block Centre
RIM	railway infrastructure manager
TPM	trusted platform module
TRL	technology readiness level
V2I	vehicle-to-infrastructure
V2V	vehicle-to-vehicle

References

1. Redaktions Netzwerk Deutschland. Ein Drittel Mehr Kapazität bei der Deutschen Bahn-so Soll es Klappen. Available online: https://www.rnd.de/wirtschaft/ein-drittel-mehr-kapazitat-bei-der-deutschen-bahn-so-soll-es-klappen-X6M3WDPHKVIEFVMDHIIOQAXOZU.html (accessed on 11 July 2023).
2. Rebhan, C. Obwohl Regierung Mehr für Klimaschutz Tun Will: Erst 2022 Gibt der Bund Mehr Geld Für Schienen Aus Als Für Straßen. Available online: https://www.businessinsider.de/politik/deutschland/obwohl-regierung-mehr-fuer-klimaschutz-tun-will-erst-2022-gibt-der-bund-mehr-geld-fuer-schienen-aus-als-fuer-strassen/ (accessed on 11 July 2023).
3. Iffländer, L.; Buder, T.; Loreth, T.; Villota, M.A.; Schmitz, W.; Neubecker, K.A.; Pickl, S. Physical Attacks on the Railway System. *arXiv* **2023**, arXiv:2306.00623.
4. Slivkova, S.; Michalcova, L. Identification and Classification of Soft Targets in Railway Infrastructure. In *Proceedings of the TRANSBALTICA XIII: Transportation Science and Technology*; Prentkovskis, O., Yatskiv (Jackiva), I., Skačkauskas, P., Maruschak, P., Karpenko, M., Eds.; Springer International Publishing: Cham, Switzerland, 2023; pp. 667–676.
5. Wang, Y.; Zhang, M.; Ma, J.; Zhou, X. Survey on Driverless Train Operation for Urban Rail Transit Systems. *Urban Rail Transit* **2016**, *2*, 106–113. https://doi.org/10.1007/s40864-016-0047-8.
6. Keevill, D. *Implications of Increasing Grade of Automation*; American Public Transportation Association: Washington, DC, USA, 2017.
7. Wunsch, S.; Lehnert, M.; Krimmling, J.; Easton, J. *Datenformate, Datenmodelle und Datenkonzepte für den Eisenbahnbetrieb*; Der Eisenbahningenieur; Eurailpress: Frankfurt am Main, Germany, 2016.
8. Schmit, M.; Kerth, S.; Sinnecker, G.; Walther, G. *Modernisierung des deutschen Eisenbahnnetzes durch Digitalisierung und ETCS-Ausrüstung*; Verband Deutscher Verkehrsunternehmen e.V. (VDV): Berlin, Germany, 2018.
9. Liu, Y.; Yuan, L. Research on Train Control System Based on Train to Train Communication. In Proceedings of the 2018 International Conference on Intelligent Rail Transportation (ICIRT), Singapore, 12–14 December 2018; pp. 1–5. https://doi.org/10.1109/ICIRT.2018.8641573.
10. Toussaint, C. *Einsatz von Drohnen im Bahnbereich*; Der Eisenbahningenieur; Eurailpress: Frankfurt am Main, Germany, 2021.
11. Schmid, G.; Sendlhofer, G.; Lexhaller, M. *Robortik im Gleisbau*; Der Eisenbahningenieur; Eurailpress: Frankfurt am Main, Germany, 2019.
12. Chew, T.; Luebkeman, C.; Morrell, M.; Goulding, L. *Future of Rail 2050*; Arup: London, UK, 2019.
13. Leining, M.; Schubert, M.; Heinrich, M.; Katzenbeisser, S.; Unger, S.; Krauß, C.; Scheuermann, D. *Prognose Securitybedarf und Bewertung möglicher Sicherheitskonzepte; Teil 1: Technologieprognose*; Deutsches Zentrum für Schienenverkehrsforschung beim Eisenbahn-Bundesamt: Dresden, Germany, 2022. https://doi.org/10.48755/dzsf.220008.06.
14. Eyben, F.; Wöllmer, M.; Poitschke, T.; Schuller, B.; Blaschke, C.; Färber, B.; Nguyen-Thien, N. Emotion on the Road: Necessity, Acceptance, and Feasibility of Affective Computing in the Car. *Adv. Hum.-Comp. Int.* **2010**, *2010*, 263593. https://doi.org/10.1155/2010/263593.
15. Costa, P.; Vasalou, A.; Pitt, J.; Dias, T.; Falcão e Cunha, J. *The Railway Blues: Affective Interaction for Personalised Transport Experiences*; ACM: New York, NY, USA, 2013. https://doi.org/10.1145/2541831.2541843.
16. Unger, S.; Arzoglou, E.; Heinrich, M.; Scheuermann, D.; Katzenbeisser, S. Risk Assessment Graphs: Utilizing Attack Graphs for Risk Assessment. *arXiv* **2023**, arXiv:2307.14114.
17. Heinrich, M.; Iffländer, L. Softwaregestützte Bedrohungsanalyse durch Angriffsgraphen. *Signal Draht* **2022**, *5*, 28–34.
18. DIN VDE V 0831-104:2015-10, Elektrische Bahn-Signalanlagen - Teil 104: Leitfaden für die IT-Sicherheit auf Grundlage IEC 62443. Available online: https://www.din.de/de/mitwirken/normenausschuesse/dke/veroeffentlichungen/wdc-beuth:din21:234969038 (accessed on 27 July 2023).
19. International Electrotechnical Commission. IEC 62443: Industrial Communication Networks—Network and System Security. Available online: https://webstore.iec.ch/preview/info_iec62443-3-3%7Bed1.0%7Den.pdf (accessed on 27 July 2023).
20. Huang, L.; Joseph, A.D.; Nelson, B.; Rubinstein, B.I.; Tygar, J.D. Adversarial Machine Learning. In Proceedings of the 4th ACM Workshop on Security and Artificial Intelligence, Chicago, IL, USA, 21 October 2011; Association for Computing Machinery: New York, NY, USA, 2011; pp. 43–58. https://doi.org/10.1145/2046684.2046692.

21. National Aeronautics and Space Administration. Technology Readiness Level. 2019. Available online: https://www.nasa.gov/directorates/heo/scan/engineering/technology/technology_readiness_level (accessed on 22 September 2023).
22. Schildt, H. *BSI IT-Grundschutz-Compendium Edition 2022*; Technical Report; Bundesamt für Sicherheit und Informationstechnik: Berlin, Germany, 2022.
23. BSI. *Konkretisierung der Anforderungen an die Gemäß § 8a Absatz 1 BSIG Umzusetzenden Maßnahmen*; Technical Report; Bundesamt für Sicherheit und Informationstechnik: Berlin, Germany, 2020.
24. Reznik, L. Malware and Vulnerabilities Detection and Protection. In *Intelligent Security Systems: How Artificial Intelligence, Machine Learning and Data Science Work For and Against Computer Security*; Wiley: Hoboken, NJ, USA, 2022; pp. 177–246. https://doi.org/10.1002/9781119771579.ch4.
25. Ali, J.A.; Nasir, Q.; Dweiri, F.T. Business Continuity Management Framework of Internet of Things (IoT). In Proceedings of the 2019 Advances in Science and Engineering Technology International Conferences (ASET), Dubai, United Arab Emirates, 26 March–10 April 2019; pp. 1–7. https://doi.org/10.1109/ICASET.2019.8714483.
26. *ISO 22301:2019*. Security and Resilience—Business Continuity Management Systems—Requirements 2019. International Organization for Standardization: Geneva, Switzerland, 2019.
27. Johari, R.; Kaur, I.; Tripathi, R.; Gupta, K. Penetration Testing in IoT Network. In Proceedings of the 2020 5th International Conference on Computing, Communication and Security (ICCCS), Patna, India, 14–16 October 2020; pp. 1–7. https://doi.org/10.1109/ICCCS49678.2020.9276853.
28. Heinrich, M.; Gölz, A.; Arul, T.; Katzenbeisser, S. Rule-based anomaly detection for railway signalling networks. *Int. J. Crit. Infrastruct. Prot.* **2023**, *42*, 100603. https://doi.org/10.1016/j.ijcip.2023.100603.
29. McParland, C.; Peisert, S.; Scaglione, A. Monitoring Security of Networked Control Systems: It's the Physics. *IEEE Secur. Priv.* **2014**, *12*, 32–39. https://doi.org/10.1109/MSP.2014.122.
30. Maffei, M. Formal Methods for the Security Analysis of Smart Contracts. In Proceedings of the 2021 Formal Methods in Computer Aided Design (FMCAD), New Haven, CT, USA, 19–22 October 2021; pp. 1–2. https://doi.org/10.34727/2021/isbn.978-3-85448-046-4_3.
31. Banks, A.S.; Kisiel, M.; Korsholm, P. Remote attestation: A literature review. *arXiv* **2021**, arXiv:2105.02466.
32. Du, C.; Dutta, S.; Kurup, P.; Yu, T.; Wang, X. A review of railway infrastructure monitoring using fiber optic sensors. *Sens. Actuators A Phys.* **2020**, *303*, 111728.
33. Maes, R.; Verbauwhede, I. Physically unclonable functions: A study on the state of the art and future research directions. In *Towards Hardware-Intrinsic Security: Foundations and Practice*; Spring: Berlin/Heidelberg, Germany, 2010; pp. 3–37.
34. Gao, Y.; Al-Sarawi, S.F.; Abbott, D. Physical unclonable functions. *Nat. Electron.* **2020**, *3*, 81–91.
35. Heinrich, M.; Vateva-Gurova, T.; Arul, T.; Katzenbeisser, S.; Suri, N.; Birkholz, H.; Fuchs, A.; Krauß, C.; Zhdanova, M.; Kuzhiyelil, D.; et al. Security Requirements Engineering in Safety-Critical Railway Signalling Networks. *Secur. Commun. Netw.* **2019**, *2019*, 1–14. https://doi.org/10.1155/2019/8348925.
36. Heinrich, M.; Renkel, D.; Arul, T.; Katzenbeisser, S. Predicting Railway Signalling Commands using Neural Networks for Anomaly Detection. In Proceedings of the International Conference on Computer Safety, Reliability, and Security, Lisbon, Portugal, 16–18 September 2020; Springer: Berlin/Heidelberg, Germany, 2020. https://doi.org/10.1007/978-3-030-54549-9_11.
37. D'Acquisto, G.; Domingo-Ferrer, J.; Kikiras, P.; Torra, V.; de Montjoye, Y.A.; Bourka, A. Privacy by design in big data: An overview of privacy enhancing technologies in the era of big data analytics. *arXiv* **2015**, arXiv:1512.06000.
38. Kaaniche, N.; Laurent, M.; Belguith, S. Privacy enhancing technologies for solving the privacy-personalization paradox: Taxonomy and survey. *J. Netw. Comput. Appl.* **2020**, *171*, 102807.
39. Heurix, J.; Zimmermann, P.; Neubauer, T.; Fenz, S. A taxonomy for privacy enhancing technologies. *Comput. Secur.* **2015**, *53*, 1–17.
40. Heydt-Benjamin, T.S.; Chae, H.J.; Defend, B.; Fu, K. Privacy for public transportation. In Proceedings of the Privacy Enhancing Technologies: 6th International Workshop, PET 2006, Cambridge, UK, 28–30 June 2006; Revised Selected Papers 6; Springer: Berlin/Heidelberg, Germany, 2006; pp. 1–19.
41. López-Aguilar, P.; Batista, E.; Martínez-Ballesté, A.; Solanas, A. Information Security and Privacy in Railway Transportation: A Systematic Review. *Sensors* **2022**, *22*, 7698.
42. Álvarez, A.; Trapero, R.; Guilhot, D.; García-Mila, I.; Hernandez, F.; Marín-Tordera, E.; Forne, J.; Masip-Bruin, X.; Suri, N.; Heinrich, M.; et al. CIPSEC-Enhancing Critical Infrastructure Protection with Innovative Security Framework. In *River Publishers Series in Security and Digital Forensics*; Challenges in Cybersecurity and Privacy–the European Research Landscape; River Publishers: Aalborg, Denmark, 2019; Chapter 7, pp. 129–148. https://doi.org/10.13052/rp-9788770220873.
43. Khan, S.A.; Malik, S.A. Adaptive beamforming algorithms for anti-jamming. *Int. J. Signal Process. Image Process. Pattern Recognit.* **2011**, *4*, 95–106.
44. Yu, K.B.; Murrow, D.J. Adaptive digital beamforming for angle estimation in jamming. *IEEE Trans. Aerosp. Electron. Syst.* **2001**, *37*, 508–523.
45. Kong, Z.; Yang, S.; Wang, D.; Hanzo, L. Robust beamforming and jamming for enhancing the physical layer security of full duplex radios. *IEEE Trans. Inf. Forensics Secur.* **2019**, *14*, 3151–3159.

46. Morgulis, N.; Kreines, A.; Mendelowitz, S.; Weisglass, Y. Fooling a real car with adversarial traffic signs. *arXiv* **2019**, arXiv:1907.00374.
47. Ntalampiras, S.; Misuraca, G.; Rossel, P. *Artificial Intelligence and Cybersecurity Research;* Technical Report; ENISA: Attiki, Greece 2023. https://doi.org/10.2824/808362.

vehicles

MDPI

Article

Ride Comfort Improvements on Disturbed Railroads Using Model Predictive Control

Alexander Posseckert * and Daniel Lüdicke

Institute of System Dynamics and Control, German Aerospace Center (DLR), Münchener Straße 20, 82234 Weßling, Germany; daniel.luedicke@dlr.de

* Correspondence: alexander.posseckert@dlr.de; Tel.: +49-8153-28-1740

Abstract: This paper proposes a control strategy for active lateral secondary suspension that uses preview data. Based on a derived analytical model, a model predictive controller (MPC) is implemented. The influence of the track irregularities upon carbody lateral dynamics is considered explicitly. The controller developed is applied to a full-scale rail vehicle model. Ride comfort is evaluated according to EN 12299. Multibody simulations show that there is a significant increase in continuous ride comfort on poor-quality tracks.

Keywords: railway vehicle; comfort; active suspension; MPC; preview; track irregularities

1. Introduction

Railway transport will play an even more important role in future mobility because of its environmental friendliness. Along with this, a further increase in reliability, capacity, and attractiveness is inevitable. DLR's Next-Generation Train (NGT) project is investigating what future high-speed trains could look like [1–3]. Its main features are a lightweight, double-deck carbody design and a mechatronic running gear with Driven Independently Rotating Wheels (DIRW).

The increasing digitalization of the transport sector supports better communication and information processing. Today's modern signalling communication between train and infrastructure is characterised by communication-based train control systems such as ERTMS/ETCS [4]. Modern passenger trains, e.g., Siemens Velaro, already use mobile communications for business-related functions such as system health status, predictive maintenance or passenger information. Ethernet-based on-train communication networks are already specified in the IEC 61375 standard [5,6] and translated into national standards, e.g., EN 61375 [7]. The growth of available bandwidth and more flexibility enable new applications. Train communication network (TCN) innovations are being developed in Shift2Rail projects [8] and will be continued in ERJU projects [9]. With the increasing connectivity of rail transport in general, the availability of information about the train and its environment is also improving accordingly. Existing information from land-side databases, cloud services or other (preceding) trains can be transferred more easily and used more widely. For example, information about a train's current and future route enables mechatronic running gears to adapt to track conditions.

The running gear serves as a link between the track and the carbody of a railway vehicle. Its suspension carries the weight of the carbody, provides movement guidance and facilitates acceleration or braking [10]. During the ride, the wheel pairs experience track excitations due to routing and track deviations. However, the running gear must minimize the acceleration and jerk experienced by passengers in order to improve vibration comfort. Various suspension technologies have been developed to decouple track excitations from the carbody movement. The suspension systems can be broadly classified into passive, semi-active and active ones. A general overview regarding this research topic is given in [11–13]. (Semi-)active suspension has an actuation or control system that relies on sensor data. With semi-active dampers, the damping rate can be adjusted and, thus, influence the dissipation of kinetic energy. Active spring-damper systems can absorb kinetic energy and also actively exert forces. Currently, passive and (semi-)active secondary suspension systems mostly aim to filter or dampen track irregularities experienced as stochastic effects in a reactive manner. The combination of new data sources enabled by digitalisation along with active suspension and innovative controllers opens up new potential for rail vehicles.

Model Predictive Control (MPC) is a promising control strategy for active suspension. Its main advantage lies in the ability to anticipate and compensate for future disturbances that impact the system. Furthermore, MPC can account for state restrictions or actuator limitations directly. Although MPC is adopted for active suspension in automotive research, e.g., [14–16], it is not commonly utilized for active suspension in railway technology. MPC is used to avoid hunting instability in railway vehicles. This is reported in [17] via the use of active yaw dampers. The use of DIRW in NGT's running gear design means that hunting instability is unlikely to occur. Another MPC-based approach was implemented in [18–20]. The studies only focus on vertical degrees of freedom, such as bouncing and pitching. Acceleration as well as suspension deflection can be successfully reduced; however, in order to evaluate comfort holistically, lateral motion has to be taken into account as well.

In order to improve ride comfort using active suspension, a detailed knowledge of track characteristics and their positions would be advantageous. The potential for improvement remains underexplored due to a lack of real-time track data, communication possibilities, and mechatronic actuators. In this regard, the present work outlines a control method for active lateral secondary suspension based on MPC. It is investigated how knowledge about the track ahead can be used for MPC in order to increase ride comfort.

Section 2 introduces the analytical model of the vehicle and the MPC control method based on it. Simulation results and the potential for improvement of the method are presented in Section 3. Section 4 provides further discussions of the results, followed by a summary in Section 5.

2. Modeling

This paper focuses on lateral dynamics; therefore, the following degrees of freedom (DOF) are considered for the carbody: Lateral displacement y_{cb}, roll angle ϕ_{cb} and yaw angle ψ_{cb}. In addition, each bogie is allowed to rotate around the x-axis by the angle $\phi_{bg,fr/re}$. This yields a model with 5 DOF, as illustrated in Figure 1. Note that a more detailed description is given in Figure A1.

$$q = \begin{bmatrix} y_{cb} & \phi_{cb} & \phi_{bg,fr} & \phi_{bg,re} & \psi_{cb} \end{bmatrix}^T. \tag{1}$$

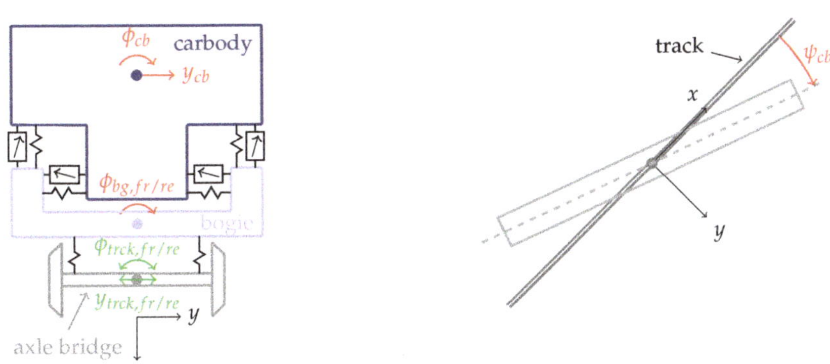

Figure 1. Definition degrees of freedom (red). Furthermore, the base excitation of the axle bridge due to track irregularities (green) is visualized. (**Left**) Cross section. (**Right**) Top view.

During the creation of the analytical model, track irregularities are considered explicitly. In order to achieve this, track errors are modelled as base excitation of the axle bridge; see Figure 1. The interaction between wheel and rail is simplified as follows. The running gear design of the NGT includes DIRW. While velocity-dependent hunting motion no longer applies, lateral guidance control is required in order to prevent flange contact; see [2]. The controller designed for this purpose significantly influences the axle bridge movement and, thus, the impact of track errors upon carbody dynamics. Figure 2 illustrates the step response of the axle bridge with respect to a lateral displacement. The result was obtained from a multibody simulation (MBS) of the NGT at a velocity of 120 $\mathrm{km\,h^{-1}}$. In general, the response depends on velocity. In order to take this complex closed-loop behavior into account when creating the model, it is proposed to approximate the resulting lateral dynamics of the axle bridge via a transfer function. The algorithm used for the determination of the transfer function parameters is based on a least squares method; see [21]. It is found that a transfer function with a numerator degree of 1 and denominator degree of 2, as given in Equation (2), is sufficient to approximate the illustrated response; see Figure 2. To conclude, the axle bridge of the analytical model is excited by track irregularities which have been filtered through a velocity dependent transfer function.

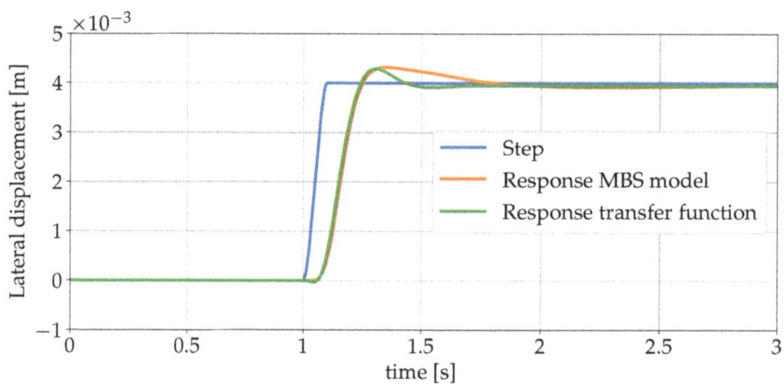

Figure 2. Step response for lateral displacement of the axle bridge. Orange: MBS model. Yellow: Derived model where the axle bride motion is simplified by a transfer function.

$$TF_{120} = \frac{-5.06\,s + 279.4}{s^2 + 20.5\,s + 283.2}. \tag{2}$$

The main focus of this work is lateral dynamics. Therefore, vertical components of track irregularities are disregarded. The remaining components of track excitation at both front and rear bogie are collected in the disturbance vector

$$z = \begin{bmatrix} \phi_{trck,fr} & \dot{\phi}_{trck,fr} & \phi_{trck,re} & \dot{\phi}_{trck,re} & y_{trck,fr} & \dot{y}_{trck,fr} & y_{trck,re} & \dot{y}_{trck,re} \end{bmatrix}^T. \tag{3}$$

While deriving the equations of motion via Newton–Euler, the following additional assumptions are made:

- Due to the mechatronic guidance of NGT's running gear, flange contact does not occur.
- Bumpstop contact can be avoided since MPC has the ability to handle state restrictions (e.g., suspension deflections).
- Small angles of roll and yaw are assumed.

Based on these assumptions, the equations of motion can be linearized. A detailed description is given in Appendix A

$$\boldsymbol{M}\ddot{\boldsymbol{q}} + \boldsymbol{D}\dot{\boldsymbol{q}} + \boldsymbol{K}\boldsymbol{q} = \boldsymbol{F}_{act}\,\boldsymbol{u} + \boldsymbol{F}_{dist}\,\boldsymbol{z}. \tag{4}$$

Subsequently, the system is transformed into state space

$$\begin{aligned} \dot{x} &= Ax + Bu + Ez \\ y &= Cx \end{aligned} \tag{5}$$

with

$$x = \begin{bmatrix} q & \dot{q} \end{bmatrix}^T. \tag{6}$$

In the secondary suspension, the actuators are mounted in both lateral and vertical directions; see Figure 1. Consequently, forces can be applied in the lateral direction as well as torques around the x and z-axis

$$u = \begin{bmatrix} F_{lat} & M_{roll} & M_{yaw} \end{bmatrix}^T. \tag{7}$$

In the disturbance vector z, the influence of track layout (e.g., centrifugal force in small radius curves) might be integrated as well; however, this paper focuses on disturbances due to track irregularities. The carbody DOF are chosen as output

$$y = \begin{bmatrix} y_{cb} & \phi_{cb} & \psi_{cb} \end{bmatrix}^T. \tag{8}$$

Equation (5) is now transformed to discrete formulation with timestep k using the first-order hold method. Furthermore, control variable u can be written via its change in one timestep and its previous value; see Equation (9). Both formulations are equivalent, though it is easier to incorporate actuator dynamics later by restricting the change in each timestep

$$u(k) = u(k-1) + \Delta u(k). \tag{9}$$

Finally, one obtains

$$\begin{aligned} x(k+1) &= A_d x(k) + B_d u(k-1) + B_d \Delta u(k) + E_d z(k) \\ y(k) &= C_d x(k). \end{aligned} \tag{10}$$

Using the model described in Equation (10), MPC minimizes a cost function J over the prediction horizon n_p by determining an optimal sequence of control variables over the control horizon n_c. Then, the first element of the sequence is applied and the process is repeated in the next timestep $k+1$, see [22]. When setting up the cost function, system output and control variables are collected in a single vector for the whole prediction horizon and control horizon, respectively

$$\bar{y}(k+1) = \begin{bmatrix} y(k+1) \\ y(k+2) \\ \vdots \\ y(k+n_p) \end{bmatrix}, \qquad \Delta\bar{u}(k) = \begin{bmatrix} \Delta u(k) \\ \Delta u(k+1) \\ \vdots \\ \Delta u(k+n_c-1) \end{bmatrix}. \tag{11}$$

The same is carried out for disturbances

$$\bar{z}(k) = \begin{bmatrix} z(k) \\ z(k+1) \\ \vdots \\ z(k+n_p-1) \end{bmatrix}. \tag{12}$$

The output prediction over the prediction horizon is given by

$$\bar{y}(k+1) = Fx(k) + Gu(k-1) + S\bar{z}(k) + H\Delta\bar{u}(k), \tag{13}$$

where $\boldsymbol{F}$, $\boldsymbol{G}$, $\boldsymbol{S}$ and $\boldsymbol{H}$ can be derived from the system matrices; see Equation (A9). With weighting matrices $\boldsymbol{Q}$ and $\boldsymbol{R}$, the cost function J that has to be minimized can be formulated as

$$\begin{aligned} \min_{\Delta\bar{\boldsymbol{u}}(k)} \quad & J = \bar{\boldsymbol{y}}(k+1)^T \boldsymbol{Q} \bar{\boldsymbol{y}}(k+1) + \Delta\bar{\boldsymbol{u}}(k)^T \boldsymbol{R} \Delta\bar{\boldsymbol{u}}(k) \\ \text{s.t.} \quad & \Delta\boldsymbol{u}_{min} \leq \Delta\boldsymbol{u}(k) \leq \Delta\boldsymbol{u}_{max} \\ & \boldsymbol{y}_{min} \leq \boldsymbol{y}(k) \leq \boldsymbol{y}_{max}. \end{aligned} \tag{14}$$

3. Simulation Results

The derived controller is applied to a full-scale vehicle model of the NGT. Due to several of its main features (e.g., lightweight carbody design in doubledeck configuration) it has turned out to be challenging to achieve a sufficient ride comfort with conventional passive secondary suspension. Therefore, the vehicle offers an ideal platform for testing the controller. The simulation consists of two main parts. Multibody simulations are performed in the Simpack 2021.3 software. For the controller, a function has been written in Simulink (Matlab) based on the MPC realization described in [23]. The optimization problem is solved using Matlab 2019b command quadprog based on an interior-point algorithm [24]. Both software products are interconnected using the co-simulation interface Simat.

A track has been chosen that is comparable to EN 14363 test zone 1 [25], see Figure A2. In this scenario, track parameters like curvature and superelevation are not expected to have a significant influence upon lateral dynamics. Lateral and crosslevel track irregularities are chosen according to ERRI B176 high [26]. The standard deviation of lateral alignment Δy^0_σ is 1.19 mm. At a speed of 120 km h^{-1}, this is within the target test range TL90 defined in EN 14363; see Table 1. For the crosslevel direction, no specifications are made in EN 14363. In addition, TL90 applies to the left and right rail separately, while ERRI B176 is for track-related excitations (i.e., for both rails at once). In order to ensure that the TL90 criterion for the longitudinal level Δz^0_σ is not exceeded, no additional vertical track irregularities are used. For the selected crosslevel excitation, the standard deviation of vertical distance between the left and right rail is 2.12 mm. Therefore, the requirements of EN 14363 regarding track geometric quality are met; see Table 1. The evaluation of ride comfort is carried out according to EN 12299 [27]. This means the variable N_{MVy} is determined, which incorporates lateral acceleration measured at several positions on both floors of the carbody; see Figure 3.

Table 1. Track geometric quality: Target test ranges for standard deviation TL90 [25].

Reference Speed	Alignment Δy^0_σ	Longitudinal Level Δz^0_σ
$V \leq 120$ km h^{-1}	1.05–1.45 mm	1.80–2.50 mm

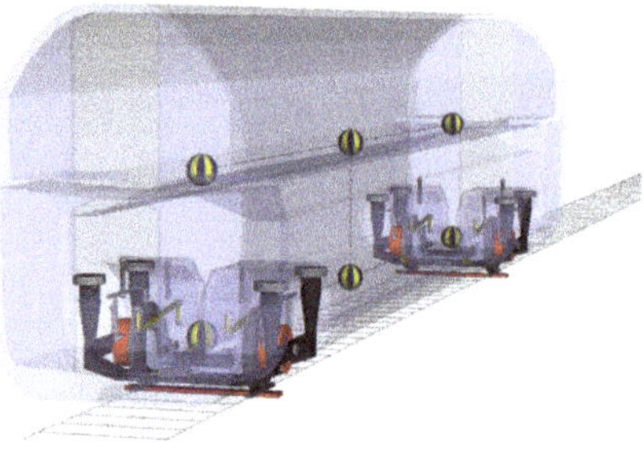

Figure 3. MBS model of a NGT middle wagon. Measurement points are visualized as yellow spheres.

Numerical results are presented in Figure 4. Note that only the lateral comfort index N_{MVy} is considered since this paper focuses on lateral dynamics. In order to evaluate ride comfort holistically, vertical and longitudinal motion has to be taken into account as well.

Compared to a model of the NGT with passive secondary suspension, a better ride comfort is achieved at all measurement points with MPC, recognizable by a lower N_{MVy} value. For instance, on the top deck at front sensor position there is an improvement of 47%. It is noticeable that at both center-measurement points, ride comfort is better compared to front and rear measurement points. At measurement point center bottom, for example, N_{MVy} value is only 30% of the one determined at rear bottom. When comparing ride comfort for the lower and upper floor, no larger differences are apparent.

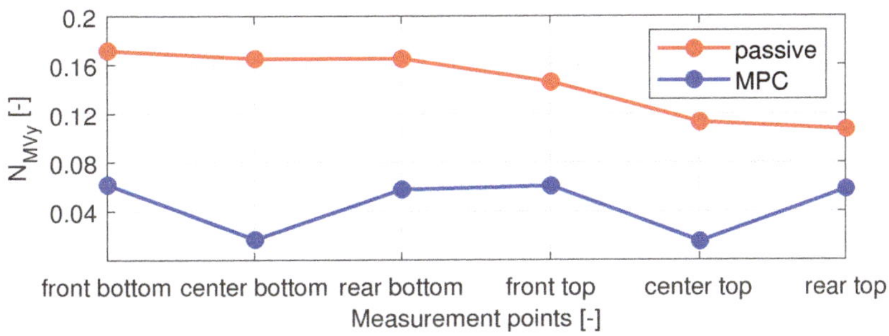

Figure 4. Lateral ride comfort index evaluated at different measurement points.

In order to predict future system behaviour, MPC requires the current train position. To obtain information regarding the required positioning accuracy, several simulations are carried out with different localization errors. For example, a test case is performed in which there is an intentional error of +2 m in positioning between the actual train position and the assumed position fed into MPC. The results are given in Table 2. Lateral ride comfort obtained at two selected measurement points is given for different localization errors. Furthermore, the result of a simulation without knowledge of track data is listed as well, i.e., $\bar{z}(k) = \mathbf{0}$. The far-right column compares test cases with localization errors to the reference case without errors. It can be observed that ride comfort deteriorates by 31% if the track ahead is not known. Nevertheless, even without track data the proposed active lateral suspension still provides better comfort than the passive system; see Figure 4. If positioning accuracy is less or equal ±1.0 m, results subjected to localization errors are better than the test case without track data. Finally, it is worth mentioning that N_{MVy} values for "positive" positioning errors are better than for the "negative" ones.

Table 2. Results for lateral comfort index when the determination of the train position is faulty.

Localization Error	NMVy Center Top	NMVy Rear Top	NMVy Rear Top Comparison to Error-Free
+2.0 m	0.038	0.086	41%
+1.0 m	0.030	0.071	16%
+0.5 m	0.021	0.063	3%
±0.0 m	0.016	0.061	Reference
−0.5 m	0.022	0.067	9%
−1.0 m	0.031	0.078	28%
−2.0 m	0.039	0.099	62%
No track data	0.029	0.080	31%

Next, the energy consumption of all actuators mounted in a coach is considered. Assuming point-to-point force elements for the actuators shown in Figure 1, the directions of relative velocity v and applied forces F coincide; see [28]. Therefore, the current power P of each force element is determined using the scalar equation

$$P = F\,v. \tag{15}$$

By integrating power over time, the energy consumed by the actuators is obtained. Since EN 12299 prescribes measurement durations of 300 s, the time period is chosen accordingly. At a speed of 120 km h^{-1}, this corresponds to a track length of 10 km. Figure 5 shows the required power of a lateral actuator during simulation with peak values of approx. 80 W. In addition, negative values of power can be observed. This offers potential for recuperation.

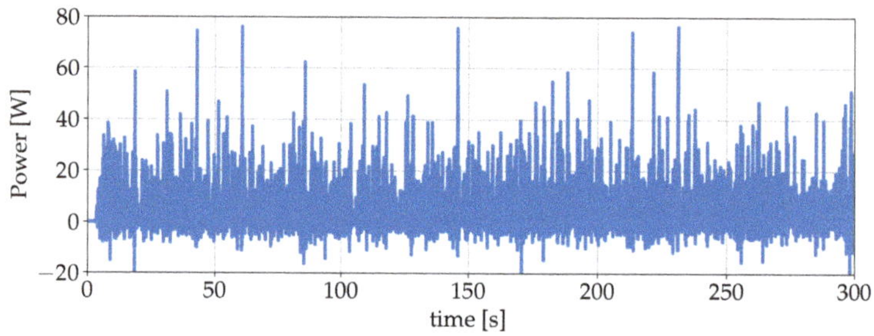

Figure 5. Power consumption of one lateral actuator during a test scenario.

The total energy summed up over all actuators of a coach is illustrated in Figure 6. For the considered track scenario, 4.5 W h of energy is consumed. For comparison, the same simulation is carried out using actuators with energy-harvesting capabilities. In this case, energy consumption is 3.9 W h, which corresponds to a reduction of 13%. Assuming an annual mileage of 500,000 km, the energy consumption for the reference case adds up to 225 kW h each year.

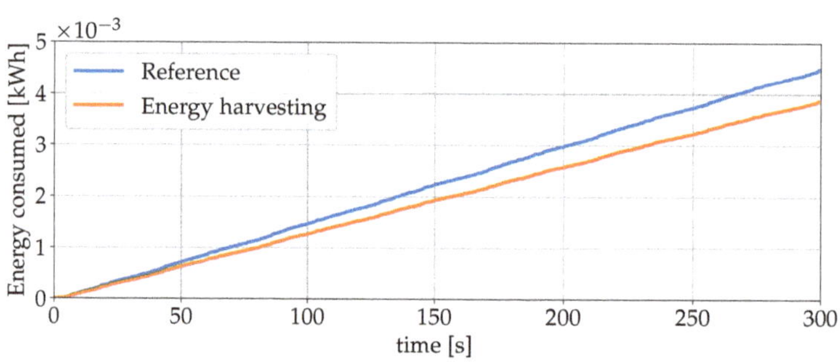

Figure 6. Consumed energy for all actuators of one coach during a test scenario. Blue: Reference case. Orange: Actuators with energy-harvesting capabilities.

4. Discussion

The results obtained in Section 3 indicate that the analytical model given in Equation (5) in combination with the assumed impact of irregularities is suitable for predicting future dynamic behaviour of the carbody. Using MPC, it was shown that, compared to a model of the NGT with passive secondary suspension, a significant increase in lateral ride comfort can be achieved; see Figure 4. It is noticeable that better N_{MVy} values are obtained near carbody's center of gravity compared to measurement points above the front and rear bogie. Consequently, ride comfort is significantly influenced by carbody yaw motion. In contrast to this, between the upper and lower floor no larger influence is detectable. Therefore, roll motion seems to be suppressed effectively.

In Table 2, the influence of positioning inaccuracy is examined. If the localization error is less or equal to ± 1 m, it is advantageous to consider the upcoming track in the control algorithm as shown in Equation (13). This behaviour is plausible since the effect of a single bump upon carbody dynamics is tied to a very specific location. Thus, if the current train position is not located precisely, the future behaviour cannot be predicted well by MPC. This represents an ambitious requirement in terms of train localisation. However, for other applications like automatic train operation, precise and reliable train position data are necessary as well. Furthermore, the controller proposed in this paper is most effective on tracks with poor track quality that are not travel at high speeds. Therefore, it is likely that the required position accuracy needed for this application can be reached. Also, it is worth mentioning that even with larger errors in localization the proposed active lateral suspension still provides better ride comfort than the passive system. It is noticeable that lateral ride comfort for "positive" localization errors is better than for the "negative" ones; see Table 2. It is assumed that this asymmetry is based on the simplified modeling of NGTs lateral guidance control.

The requirements regarding power and energy usage of the actuators are quantified in Figures 5 and 6, respectively. The energy consumption required in order to counteract track irregularities is rather low compared to other consumers. However, especially on curvy tracks with high cant deficiency, lateral centering is a relevant part of active secondary suspension; see [29]. In order to investigate energy demand holistically, both needs to be considered together. Another important aspect is the requirements regarding actuator dynamics that must be fulfilled. In Figure 7, Fourier transform is applied to a signal that represents the force of a lateral actuator. Forces up to a frequency of 10 Hz can be observed. At a speed of 120 $\mathrm{km\,h^{-1}}$, track excitation defined by ERRI high lies in this frequency range. While this was to be expected, it poses a challenging demand for the actuators. For example, in their studies on active secondary suspension, Orvnäs et al. used an electro-hydraulic actuator that only performed well up to 6 Hz; see [30]. However, the main goal of this paper is to provide a functional demonstration of MPC. Its suitability as a ride comfort controller has been proven. Lowering the required frequency range of the actuators by reducing controller gains or switching from active to semi-active suspension is an important issue which will be investigated further.

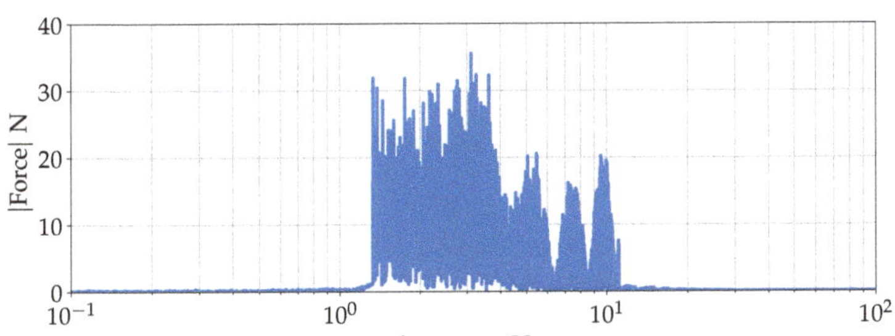

Figure 7. Force spectrum of one lateral actuator.

This paper focuses on minimizing the influence of track irregularities upon ride comfort. Using MPC, good results were obtained without considering wheel/rail-contact explicitly. Another important aspect is the influence of track layout upon carbody dynamics. It is expected that on tracks with small radius curves and high cant deficiency, there will be a non-negligible impact on ride comfort. In Section 2, it was mentioned that in the disturbance vector z, the influence of track layout (e.g., centrifugal force in small radius curves) might be integrated as well. To fully exploit the potential of MPC, both track irregularities and track layout should be considered.

5. Conclusions

A controller for active lateral secondary suspension based on Model Predictive Control (MPC) has been introduced in this paper. With the increase in digitalisation, the availability of information about the train and its environment is improving. This enables the opportunity to increase ride comfort by using information about the preceding track. To design the controller, an analytical model which describes the carbody lateral dynamics is developed. The impact of track irregularities (in particular lateral and roll excitation) is considered explicitly. The controller is applied to a multibody simulation of the NGT on the tangent track and very large radius curves according to EN 14363. On the selected tracks, there is a significant improvement of ride comfort at each measurement point.

It was found that if the localisation error is less or equal to ± 1 m, it is advantageous to consider the upcoming track in the control algorithm. Since other applications like automatic train operation require precise and reliable train position data as well, it is expected that this accuracy can be reached. Furthermore, even without track data, the proposed active lateral suspension still provides better comfort than a comparable vehicle with passive secondary suspension.

The energy consumption of the actuators for the scenario considered is rather low. However, especially on curvy tracks with high cant deficiency, energy usage might increase. In order to investigate energy demand holistically, both scenarios need to be considered together. Furthermore, it is worth mentioning that the control method proposed in this paper puts high requirements on the actuator dynamics up to 10 Hz.

Multibody simulations of the NGT showed promising results. The next step is to incorporate a more detailed model of the actuators. Furthermore, an analysis of robustness of the controller is still pending. However, it is important to emphasize that a significant part of external disturbances that might affect the system are already considered in the disturbance vector z.

Author Contributions: Conceptualization, A.P. and D.L.; methodology, A.P.; software, A.P. and D.L.; validation, A.P.; formal analysis, A.P.; investigation, A.P.; resources, A.P. and D.L.; data curation, A.P. and D.L.; writing—original draft preparation, A.P.; writing—review and editing, A.P. and D.L.; visualization, A.P.; supervision, A.P.; project administration, A.P. All authors have read and agreed to the published version of the manuscript.

Funding: This research received no external funding.

Institutional Review Board Statement: Not applicable.

Data Availability Statement: Not applicable.

Conflicts of Interest: The authors declare no conflict of interest.

Abbreviations

The following abbreviations are used in this manuscript:

NGT	Next-Generation Train.
TCN	Train communication network.
MPC	Model Predictive Control.
DIRW	Driven independently rotating wheels.
DOF	Degree of freedom.
w.r.t.	With respect to.
MBS	Multibody simulation
dist.	Distance
CoG	Center of gravity

Appendix A

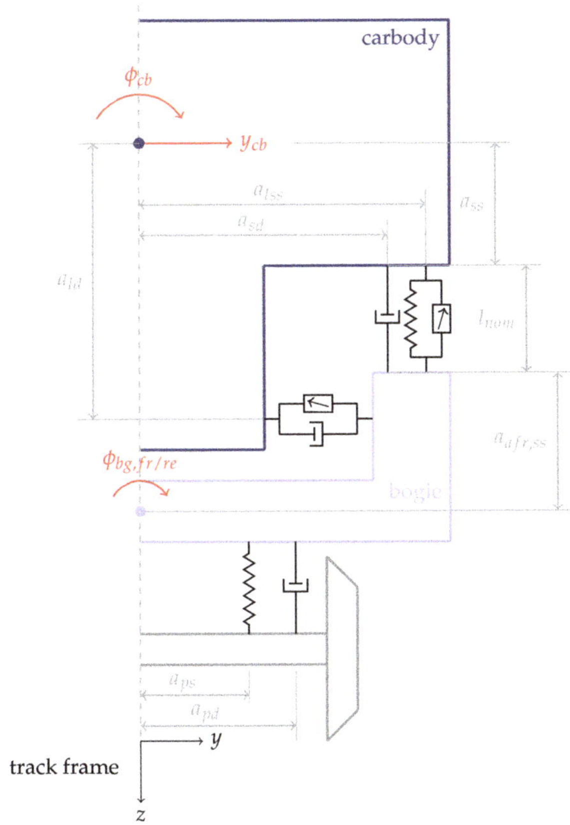

Figure A1. Detailed description of the derived model.

$$\boldsymbol{M} = \begin{bmatrix} m_{cb} & 0 & 0 & 0 & 0 \\ & I_{xx,cb} & 0 & 0 & 0 \\ & & I_{xx,bg} & 0 & 0 \\ \text{Sym.} & & & I_{xx,bg} & 0 \\ & & & & I_{zz,cb} \end{bmatrix} \tag{A1}$$

$$\boldsymbol{D} = \begin{bmatrix} d_{1,1} & d_{1,2} & 0 & 0 & 0 \\ & d_{2,2} & d_{2,3} & d_{2,4} & 0 \\ & & d_{3,3} & 0 & 0 \\ \text{Sym.} & & & d_{4,4} & 0 \\ & & & & d_{5,5} \end{bmatrix} \tag{A2}$$

$$\begin{aligned}
d_{1,1} &= 4\,d_{ld} \\
d_{1,2} &= -\ 4\,d_{ld}a_{ld} \\
d_{2,2} &= 4\,a_{ld}^2\,d_{ld} + 8\,a_{sd}^2\,d_{sd} \\
d_{2,3} &= -\ 4\,a_{sd}^2\,d_{sd} \\
d_{2,4} &= -\ 4\,a_{sd}^2\,d_{sd} \\
d_{3,3} &= 4\,a_{pd}^2\,d_{pd} + 2\,a_{ps}^2\,d_{z,ps} + 4\,a_{sd}^2\,d_{sd} \\
d_{4,4} &= 4\,a_{pd}^2\,d_{pd} + 2\,a_{ps}^2\,d_{z,ps} + 4\,a_{sd}^2\,d_{sd} \\
d_{5,5} &= 4\,d_{ld}\,a_{spd}^2
\end{aligned} \tag{A3}$$

$$\boldsymbol{K} = \begin{bmatrix} k_{1,1} & k_{1,2} & k_{1,3} & k_{1,4} & 0 \\ & k_{2,2} & k_{2,3} & k_{2,4} & 0 \\ & & k_{3,3} & 0 & k_{3,5} \\ \text{Sym.} & & & k_{4,4} & k_{4,5} \\ & & & & k_{5,5} \end{bmatrix} \tag{A4}$$

$$\begin{aligned}
k_{1,1} &= 8\,c_y \\
k_{1,2} &= -\ 8\,a_{ss}c_y - 8\,c_{y\phi} \\
k_{1,3} &= -\ 4\,a_{afr,ss}\,c_y - 4\,c_{y\phi} \\
k_{1,4} &= -\ 4\,a_{afr,ss}\,c_y - 4\,c_{y\phi} \\
k_{2,2} &= 8\,a_{lss}^2\,c_z + 8\,a_{ss}^2\,c_y - 8\,F_{pre}\,a_{ss} + 16\,a_{ss}\,c_{y\phi} + 8\,c_\phi \\
k_{2,3} &= 4\,a_{afr,ss}\,a_{ss}\,c_y - 4\,a_{lss}^2\,c_z + 4\,a_{afr,ss}\,c_{y\phi} + 4\,a_{ss}\,c_{y\phi} + 4\,l_{nom}\,c_{y\phi} - 4\,c_\phi \\
k_{2,4} &= 4\,a_{afr,ss}\,a_{ss}\,c_y - 4\,a_{lss}^2\,c_z + 4\,a_{afr,ss}\,c_{y\phi} + 4\,a_{ss}\,c_{y\phi} + 4\,l_{nom}\,c_{y\phi} - 4\,c_\phi \\
k_{3,3} &= 4\,a_{afr,ss}^2\,c_y + 4\,a_{lss}^2\,c_z + 2\,a_{ps}^2\,c_{z,ps} - 4\,F_{pre}\,a_{afr,ss} + 8\,a_{afr,ss}\,c_{y\phi} + 4\,c_\phi \\
k_{3,5} &= -\ 4\,a_{afr,ss}\,a_{spd}\,c_y - 4\,a_{spd}\,c_{y\phi} \\
k_{4,4} &= 4\,a_{afr,ss}^2\,c_y + 4\,a_{lss}^2\,c_z + 2\,a_{ps}^2\,c_{z,ps} - 4\,F_{pre}\,a_{afr,ss} + 8\,a_{afr,ss}\,c_{y\phi} + 4\,c_\phi \\
k_{4,5} &= 4\,a_{afr,ss}\,a_{spd}\,c_y + 4\,a_{spd}\,c_{y\phi} \\
k_{5,5} &= 8\,c_y\,a_{spd}^2
\end{aligned} \tag{A5}$$

$$\boldsymbol{F}_{act} = \begin{bmatrix} -1 & 0 & 0 \\ a_{ld} & 1 & 0 \\ 0 & -0.5 & 0 \\ 0 & -0.5 & 0 \\ 0 & 0 & 1 \end{bmatrix} \tag{A6}$$

$$\boldsymbol{F}_{dist} = \begin{bmatrix} 0 & 0 & 0 & 0 & f_{1,5} & f_{1,6} & f_{1,7} & f_{1,8} \\ 0 & 0 & 0 & 0 & f_{2,5} & f_{2,6} & f_{2,7} & f_{2,8} \\ f_{3,1} & f_{3,2} & 0 & 0 & f_{3,5} & 0 & 0 & 0 \\ 0 & 0 & f_{4,3} & f_{4,4} & 0 & 0 & f_{4,7} & 0 \\ 0 & 0 & 0 & 0 & f_{5,5} & f_{5,6} & f_{5,7} & f_{5,8} \end{bmatrix} \tag{A7}$$

$$\begin{aligned}
&f_{1,5} = 4\,c_y, && f_{1,6} = 2\,d_{ld} && f_{1,7} = 4\,c_y && f_{1,8} = 2\,d_{ld} \\
&f_{2,5} = -4\,c_y\,a_{ss} - 4\,c_{y\phi} && f_{2,6} = -2\,d_{ld}\,a_{ld} && f_{2,7} = -4\,c_y\,a_{ss} - 4\,c_{y\phi} && f_{2,8} = -2\,d_{ld}\,a_{ld} \\
&f_{3,1} = 2\,c_{z,ps}\,a_{ps}^2 && f_{3,2} = 4\,d_{pd}\,a_{pd}^2 + 2\,d_{z,ps}\,a_{ps}^2 && f_{3,5} = -4\,c_y\,a_{afr,ss} - 4\,c_{y\phi} && \\
&f_{4,3} = 2\,c_{z,ps}\,a_{ps}^2 && f_{4,4} = 4\,d_{pd}\,a_{pd}^2 + 2\,d_{z,ps}\,a_{ps}^2 && f_{4,7} = -4\,c_y\,a_{afr,ss} - 4\,c_{y\phi} && \\
&f_{5,5} = 4\,c_y\,a_{spd} && f_{5,6} = 2\,d_{ld}\,a_{spd} && f_{5,7} = -4\,c_y\,a_{spd} && f_{5,8} = -2\,d_{ld}\,a_{spd}
\end{aligned} \tag{A8}$$

Table A1. Name and description of parameters that are used in connection with the analytical model.

Name	Description
m_{cb}	Carbody mass
$I_{xx,cb}$	Moment of inertia around x-axis, carbody
$I_{zz,cb}$	Moment of inertia around z-axis, carbody
$I_{xx,bg}$	Moment of inertia around x-axis, bogie
c_y	Lateral stiffness secondary suspension
c_z	Vertical stiffness secondary suspension
c_ϕ	Bending stiffness secondary suspension
$c_{y\phi}$	Cross-coupling stiffness secondary suspension
$c_{z,ps}$	Vertical stiffness primary suspension
d_{ld}	Lateral damping secondary suspension
d_{sd}	Vertical damping secondary suspension
d_{pd}	Vertical damping primary dampers
d_{ps}	Vertical damping primary suspension
a_{lss}	Half the distance between secondary springs
a_{sd}	Half the distance between secondary dampers
a_{ps}	Half the distance between primary springs
a_{pd}	Half the distance between primary dampers
a_{spd}	Half the distance between the center pivots
a_{ld}	Vertical dist. lateral damper to carbody CoG
a_{ss}	Vertical dist. secondary spring to carbody CoG
$a_{afr,ss}$	Vertical dist. bogie CoG to secondary spring
l_{nom}	Nominal length secondary springs
F_{pre}	Preload of secondary springs due to carbody mass

Appendix B

$$F = \begin{bmatrix} CA \\ CA^2 \\ CA^3 \\ \vdots \\ CA^{n_p} \end{bmatrix}, \quad G = \begin{bmatrix} CB \\ C(A+I)B \\ C(A^2+A+I)B \\ \vdots \\ C\left(A^{n_p-1}+\ldots+I\right)B \end{bmatrix}$$

$$H = \begin{bmatrix} CB & 0 & \cdots & 0 \\ C(A+I)B & CB & \cdots & 0 \\ C(A^2+A+I)B & C(A+I)B & \cdots & 0 \\ \vdots & \vdots & \ddots & \vdots \\ C\left(A^{n_c-1}+\ldots+I\right)B & C(A^{n_c-2}+\ldots+I)B & \cdots & CB \\ C(A^{n_c}+\ldots+I)B & C\left(A^{n_c-1}+\ldots+I\right)B & \cdots & C(A+I)B \\ \vdots & \vdots & \ddots & \vdots \\ C\left(A^{n_p-1}+\ldots+I\right)B & C(A^{n_p-2}+\ldots+I)B & \cdots & C(A^{n_p-n_c}+\ldots+I)B \end{bmatrix} \tag{A9}$$

$$S = \begin{bmatrix} CE & 0 & 0 & \cdots & 0 \\ CAE & CE & 0 & \cdots & 0 \\ CA^2E & CAE & CE & \cdots & 0 \\ \vdots & \vdots & \vdots & \ddots & \vdots \\ CA^{n_p-1}E & CA^{n_p-2}E & CA^{n_p-3}E & \cdots & CE \end{bmatrix}$$

Appendix C

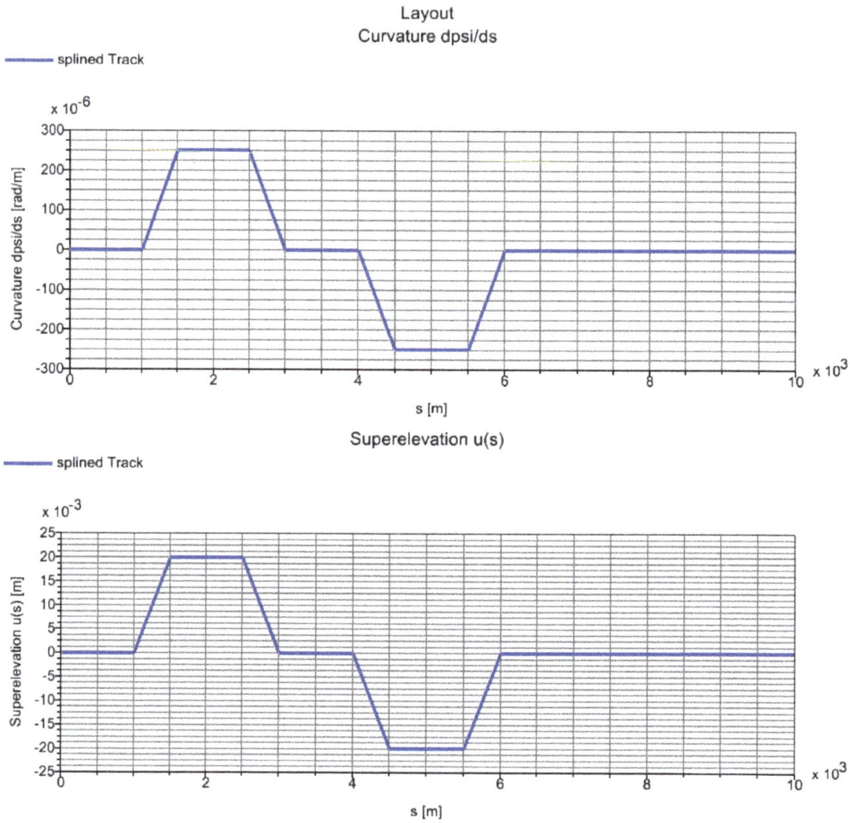

Figure A2. Track scenario.

References

1. Winter, J. Novel Rail Vehicle Concepts for a High Speed Train: The Next Generation Train. In Proceedings of the First International Conference on Railway Technology: Research, Development and Maintenance, Las Palmas de Gran Canaria, Spain, 18–20 April 2012; Civil-Comp Proceedings; Pombo, J., Ed.; Civil-Comp Press: Stirlingshire, UK, 2012. [CrossRef]
2. Heckmann, A.; Keck, A.; Grether, G. Active Guidance of a Railway Running Gear with Independently Rotating Wheels. In Proceedings of the 2020 IEEE Vehicle Power and Propulsion Conference (VPPC), Piscataway, NJ, USA, 18 November–16 December 2020; pp. 1–5. [CrossRef]
3. Heckmann, A.; Lüdicke, D.; Keck, A.; Goetjes, B. A Research Facility for the Next Generation Train Running Gear in True Scale. In *Advances in Dynamics of Vehicles on Roads and Tracks II*; Lecture Notes in Mechanical Engineering; Orlova, A., Cole, D., Eds.; Springer International Publishing: Cham, Switzerland, 2022; Volume 9, pp. 18–27. [CrossRef]
4. European Union Agency for Railways. Technical Specifications for Interoperability—Control Command and Signalling. Technical Report CCS. Available online: https://www.era.europa.eu/domains/technical-specifications-interoperability/control-command-and-signalling-tsi_en (accessed on 27 May 2016).
5. Ludicke, D.; Lehner, A. Train Communication Networks and Prospects. *IEEE Commun. Mag.* **2019**, *57*, 39–43. [CrossRef]
6. International Electrotechnical Commission. *Electronic Railway Equipment—Train communication Network (TCN)—Part 1: General Architecture*. Technical Report 61375-1; International Electrotechnical Commission: Geneva, Switzerland, 2012.
7. *DIN EN 61375-1:2015-02, Elektronische Betriebsmittel für Bahnen—Zug-Kommunikations-Netzwerk (TCN)—Teil 1: Allgemeiner Aufbau (IEC 61375-1:2012); Deutsche Fassung EN 61375-1:2012*; Technical Report; Deutsches Institut für Normung e.V.: Berlin, Germany, 2015. [CrossRef]
8. Shift2Rail CONNECTA. Final Report on the Contribution of CONNECTA to Shift2Rail: Deliverable D8.3. Available online: https://projects.shift2rail.org/download.aspx?id=e489bd37-4ec1-4fbf-bd1d-818bb0fc0762 (accessed on 14 August 2023).
9. Europe's Rail. Home—Europe's Rail. Available online: https://rail-research.europa.eu/ (accessed on 14 August 2023).

10. Schindler, C.; Brandhorst, M.; Dellmann, T.; Haigermoser, A.; Hecht, M.; Karch, S.; Löffler, G.; Rösch, W. (Eds.) *Handbuch Schienenfahrzeuge: Entwicklung, Produktion, Instandhaltung*, 1. auflage ed.; Eurail Press: Hamburg, Germany, 2014.
11. Fu, B.; Giossi, R.L.; Persson, R.; Stichel, S.; Bruni, S.; Goodall, R. Active suspension in railway vehicles: A literature survey. *Railw. Eng. Sci.* **2020**, *28*, 3–35. [CrossRef]
12. Qazizadeh, A. On Active Suspension in Rail Vehicles. Ph.D. Thesis, KTH Royal Institute of Technology, Stockholm, Sweden, 2017.
13. Orvnäs, A. *Methods for Reducing Vertical Carbody Vibrations of a Rail Vehicle: A literature Survey*; Järnvägsgruppen, KTH Railway Group: Stockholm, Sweden, 2010.
14. Gohrle, C.; Wagner, A.; Schindler, A.; Sawodny, O. Active suspension controller using MPC based on a full-car model with preview information. In Proceedings of the 2012 American Control Conference (ACC), Montreal, QC, Canada, 27–29 June 2012; pp. 497–502. [CrossRef]
15. Durmaz, B.E.; Kaçmaz, B.; Mutlu, İ.; Söylemez, M.T. Implementation and comparison of LQR-MPC on active suspension system. In Proceedings of the 2017 10th International Conference on Electrical and Electronics Engineering (ELECO), Bursa, Turkey, 30 November–2 December 2017; pp. 828–835.
16. Cho, B.K. Active suspension controller design using MPC with preview information. *KSME Int. J.* **1999**, *13*, 168–174. [CrossRef]
17. Chen, Y.; Wang, K.; Wang, Y.; Li, G. Model predictive control of active yaw damper for high-speed trains. In Proceedings of the Sixth International Conference on Traffic Engineering and Transportation System (ICTETS 2022), Guangzhou, China, 23–25 September 2022; Proceedings of SPIE; Zhou, J., Sheng, J., Eds.; SPIE: Bellingham, WA, USA, 2023; p. 140. [CrossRef]
18. Orukpe, P.E.; Zheng, X.; Jaimoukha, I.M.; Zolotas, A.C.; Goodall, R.M. Model predictive control based on mixed H_2/H_∞ control approach for active vibration control of railway vehicles. *Veh. Syst. Dyn.* **2008**, *46*, 151–160. [CrossRef]
19. Orukpe, P.E. Model Predictive Control Application to Flexible-Bodied Railway Vehicles for Vibration Suppression. *Int. J. Eng. Res. Afr.* **2013**, *10*, 25–35. [CrossRef]
20. Ulum, Z.; Affaf, M.; Salmah; Suparwanto, A. Active suspension systems design of a light rail vehicle using MPC with preview information disturbance. In Proceedings of the 2017 5th International Conference on Instrumentation, Control, and Automation (ICA), Yogyakarta, Indonesia, 9–11 August 2017; pp. 18–23. [CrossRef]
21. Ljung, L. *System Identification: Theory for the User*, 9th ed.; Prentice-Hall PTR information and system sciences series; Prentice-Hall PTR: Upper Saddle River, NJ, USA, 1996.
22. Borrelli, F.; Bemporad, A.; Morari, M. *Predictive Control for Linear and Hybrid Systems*; Cambridge University Press: Cambridge, UK; New York, NY, USA; Port Melbourne, Australia, 2017.
23. Adamy, J. *Nichtlineare Regelungen*; Springer: Berlin/Heidelberg, Germany, 2009. [CrossRef]
24. Coleman, T.F.; Li, Y. A Reflective Newton Method for Minimizing a Quadratic Function Subject to Bounds on Some of the Variables. *SIAM J. Optim.* **1996**, *6*, 1040–1058. [CrossRef]
25. *DIN EN 14363:2022-10, Bahnanwendungen—Versuche und Simulationen für die Zulassung der fahrtechnischen Eigenschaften von Eisenbahnfahrzeugen—Fahrverhalten und stationäre Versuche; Deutsche Fassung EN 14363:2016+A1:2018+A2:2022;* Technical Report; Deutsches Institut für Normung e.V.: Berlin, Germany, 2022. [CrossRef]
26. *ORE B 176. Bogies with steered or steering wheelsets. Report No. 1: Specifications and Preliminary Studies, Vol. 2. Specification for a Bogie with Improved Curving Characteristics;* Technical Report; Office for Research and Experiments (ORE): Utrecht, The Netherlands, 1989.
27. *DIN EN 12299:2009-08, Bahnanwendungen—Fahrkomfort für Fahrgäste; Deutsche Fassung EN 12299:2009*; Technical Report; Deutsches Institut für Normung e.V.: Berlin, Germany, 2009. [CrossRef]
28. Rill, G.; Schaeffer, T.; Borchsenius, F. *Grundlagen und computergerechte Methodik der Mehrkörpersimulation: Vertieft in Matlab-Beispielen, Übungen und Anwendungen*, 4. auflage ed.; Springer eBook Collection; Springer Vieweg: Wiesbaden/Heidelberg, Germany, 2020. [CrossRef]
29. Posseckert, A.; Lüdicke, D. Ride comfort improvements in switches using active secondary suspension with preview. In Proceedings of the Fifth International Conference on Railway Technology: Research, Development and Maintenance, Montpellier, France, 22–25 August 2022; Civil-Comp Conferences; Pombo, J., Ed.; Civil-Comp Press: Edinburgh, UK, 2023; pp. 1–6. [CrossRef]
30. Orvnäs, A.; Stichel, S.; Persson, R. Active lateral secondary suspension with H_∞ control to improve ride comfort: Simulations on a full-scale model. *Veh. Syst. Dyn.* **2011**, *49*, 1409–1422. [CrossRef]

vehicles

MDPI

Article

Modernization of Railway Wagons for Customer Satisfaction and Safety

Eva Nedeliaková, Michal Valla * and Matej Masár

Department of Railway Transport, University of Žilina, Univerzitná 1, 010 26 Žilina, Slovakia; eva.nedeliakova@uniza.sk (E.N.); matej.masar@uniza.sk (M.M.)
* Correspondence: valla@stud.uniza.sk; Tel.: +421-908687559

Abstract: This article focuses on the assessment of the efficiency of the modernization of railway wagons intended for passenger transport. The modernization of railway wagons is an important step towards ensuring the safer and more efficient operation of railways. However, this modernization is not without risks and, in addition to its many positive effects, it can also bring new challenges in the form of breakdowns and technical problems. Failures in passenger rail vehicles are a challenge facing manufacturers and transport operators, and identifying the relationships between these failures can help to improve the design, manufacture, and maintenance of these vehicles. Passenger rail vehicle failures can vary in nature, whether mechanical, electrical, or inside the vehicle. This paper attempts to provide a comprehensive view of the effectiveness of passenger rail car retrofitting based on the data collected and statistical analyses. The article further focuses on the description of selected rail passenger wagons, describes the analyses of the statistical data using the correlation between the selected rail wagon failures and their significance, in addition to whether the impact of modernization reduces the number of rail wagon failures. The results of this statistical analysis can be used to better understand the impact of modernization on different aspects of passenger transport and will help in making future investment and policy decisions in this area.

Keywords: railway passenger transport; modernization; wagons

1. Introduction

Rail passenger wagons are a key part of rail operations, having an irreplaceable function in both national and international rail passenger transport. Wagons are designed to provide safety and comfort for passengers. However, regular inspection and maintenance may not prevent faults and defects arising from its operation. In the current era, where the emphasis is on sustainability, the modernization of passenger wagons is also a means of reducing emissions and increasing the energy efficiency of rail operations. The modernization of passenger wagons is a key factor in ensuring the competitiveness of rail transport. The aim of this paper is to provide an overview of the importance and benefits of passenger car fleet modernization and to stimulate debate on the future of rail transport in an era of rapid technological change and increasing passenger demands by testing statistical hypotheses and correlations. A variety of analyses and data collection can be performed to examine statistical correlations in the context of railcar modernization, which could provide valuable insights into the relationships between various variables and the effects of modernization measures [1,2].

To perform these analyses, relevant data and statistical tools such as correlation coefficients, regression analyses, and analysis of variance are needed. The results of these analyses could help to better understand the impact of railcar retrofitting on different aspects, which could lead to informed decisions on rail operations, finances, and sustainability. For the application of statistical correlation, two series of railcars were selected, a Bmz series wagon and a Bdteer series wagon. The Bdteer passenger coaches were manufactured and put into service between 1990 and 1992; thus, they have been in operation for almost 26 years. As part of the modernization of the passenger wagons, solutions are used as far as possible to help unify the transport operator's fleet [3–5]. Bmz wagons have been converted from Bmeer wagons. A major advance in technical specification compared to the old wagons has been made in the areas of the pipe brake accelerator, which has increased braking performance, the boarding doors and front doors, which ensure greater passenger safety, and the information system, which provides external diagnostics of the wagon [6].

2. Research Design

The data were collected between April 2018 and March 2023 by the staff of the transport company as well as by the authors of the present paper. By summarizing and stratifying the data, an extensive database was created, which is divided into three parts [7]. The first part—data from the field—consists of data regarding the date of the fault detection, the location of the fault detection, the number of the wagon on which the fault occurred, the transport company owning the wagon, the category of the wagon, the structural unit on which the fault occurred, and the description of the fault. The second part—modernization—contains the wagon number, wagon series, year of modernization, and year of manufacture of the wagon. The third part—summary operational statistics—contains summary statistics obtained from the statistical yearbook of the carrier, the Statistical Office of the Slovak Republic and the Ministry of Transport [8,9]. Statistical correlation is an analytical method used to measure and express the degree of mutual relationship or association between two variables. This method is valuable for identifying how changes in one variable may reflect changes in another variable. Statistical correlation is widely employed in science, economics, sociology, and various other disciplines to examine relationships between different data or variables.

2.1. Disorders

A total of 78 types of disturbances were identified and listed in the database. Based on the assessment of the significance of the statistical feature variations, nine were selected as having the greatest impact on customer satisfaction and wagon operability—I1: doors and CZD, I2: toilets, I3: blocks. I1. WC; I4: air conditioning; I5: interior issues; I6: heating, I7: train radio, I8: lighting, I9: power distribution issues [10].

2.2. Structural Units

The database lists a total of 19 distinct structural units on which a fault has manifested itself. In terms of the significance of the variables of the considered statistical feature, five structural units were selected on the basis of the most frequent failures: C1: WC and washroom, C2: heating and air conditioning, C3: electrical equipment and wiring, C4: wagon cabinet, C5: interior fittings [11].

3. Results

Formulation of statistical hypotheses

From the point of view of the modernization of the units carried out on the passenger cars, it can be assumed that there is no statistical dependence between the individual failures and the cars. Based on the above, the statistical hypothesis H1 was formulated:

H1: *There is no significant statistical dependence between the individual failures.*

The structural units are upgraded based on the decision of the passenger wagon operator. This is mostly based on the frequency of failures. Therefore, repairing the most frequently failed unit should statistically significantly reduce the occurrence of a failure on that unit.

H2: *There is a significant statistical relationship between structural units and failures.*

The transport company regularly upgrades passenger wagons in order to increase passenger comfort but also to eliminate faults. By carrying out modernization it is possible to reduce the failure rate on passenger wagons.

H3: *The link between modernization and the reduction in breakdowns on passenger wagons is proven.*

3.1. There Is No Significant Statistical Dependence between the Different Disorders

A total of 275,895 failures were identified in the reporting period. From April to December 2018, 51,632 faults were identified. In 2019 and 2020, a total of 7027 and 42,178 faults were identified, respectively. Lastly, in 2021 and 2022, 44,433 and 56,640 failures were identified, respectively, and from January 2023 to March 2023, 10,739 failures were identified by the transport company's staff and employees. The following disorders subsequently led to the creation of a statistical correlation, the values of which are displayed in Table 1 [12,13].

Table 1. Correlation between selected passenger car failures.

	I1	I2	I3	I4	I5	I6	I7	I8	I9
I1	1								
I2	0.8569	1							
I3	0.1299	0.8934	1						
I4	0.0233	0.1093	0.0594	1					
I5	0.0834	0.1009	0.0837	0.2153	1				
I6	0.1573	0.2115	0.1080	0.7049	0.0655	1			
I7	0.0278	0.0819	0.4621	0.0255	0.0112	0.2115	1		
I8	0.3951	0.0482	0.3015	0.2145	0.2066	0.0985	0.1554	1	
I9	0.4587	0.2831	0.3591	0.1673	0.1862	0.2931	0.1576	0.9251	1

The correlation showed a statistical significance between parameters I1 and I2, I2 and I3, and I8 and I9. The results demonstrate statistical significance between the variations in the statistical feature of the disorder. A very strong correlation is shown between lighting and electrical wiring, strong between door and CZD and toilet, and between toilet and blocked vacuum toilet.

Statistically significant parameters need to be tested by a statistical significance test at alpha (0.05) level. Three pairs I1–I2, I2–I3, I8–I9 were tested.

For I1–I2, for the selected defining factor at the specified alpha confidence level, a p value of (0.0832) was obtained. Since the p value of the pair > p value is established, the dependence cannot be evaluated as reliable; i.e., we do not assume that dependence occurs in this pair.

For I2–I3, for the selected defining factor at the specified alpha confidence level, a p value of (0.1027) was obtained. Since the p value of the pair > p value is established, the dependence cannot be evaluated as reliable; i.e., we do not assume that dependence occurs in this pair.

For I8–I9, for the chosen defining factor at the established alpha confidence level, the p value reliability came out to be (0.0374). Since the p value of the pair < p value determined, it can be evaluated as reliable; i.e., we assume that dependence occurs in this pair [14–16].

The above hypothesis cannot be confirmed since a dependence between lighting and electrical wiring was found.

3.2. There Is a Significant Statistical Dependence between Structural Units and Failures

The carrier's dedicated passenger vehicles contain 19 structural units on which a failure was identified during the time period under review. By criterion selection of statistical feature variables, the structural units that are perceived by the customer—passenger the most were selected. These are five structural units: C1: toilet and washroom, C2:

heating and air conditioning, C3: electrical equipment and wiring, C4: wagon cabinet, C5: interior fittings.

Since the stated hypothesis is to find only those dependencies that are statistically significant, only dependencies where the coefficient of dependence β > 0.85 were considered. Based on the results presented in Table 2, a dependence between the following traits can be established:

- C1–I2;
- C2–(I4 + I6);
- C3–(I7 + I8 + I9);
- C4–I4;
- C5–I9.

Table 2. Correlation between selected passenger car failures and car structural units.

	I1	I2	I3	I4	I5	I6	I7	I8	I9
C1	0.7931	**0.8804**	0.0270	0.0057	0.3058	0.0991	0.0054	0.0040	0.0004
C2	0.2597	0.0434	0.1824	0.9053	0.1571	**0.9182**	0.1246	0.0806	0.7261
C3	0.2326	0.2775	0.2427	0.0330	0.7865	0.7770	**0.8649**	**0.9243**	**0.9535**
C4	0.1246	0.2191	0.0271	**0.9053**	0.1246	0.2144	0.0434	0.2053	0.0312
C5	0.0262	0.2155	0.2438	0.1747	0.1571	0.2675	0.2597	0.0259	**0.9254**

The above variations in the statistical features C and I that were strongly correlated with each other were tested by significance test at the chosen alpha level (0.05). Three pairs, one triplet, and one quadruplet were tested.

For C1–I2 (dependence of toilet and washroom design unit on the failure labeled toilet), for the selected defining factor at the specified alpha confidence level, a p value reliability value of (0.0332) was obtained. Since the p value of the pair < p value determined, the dependence can be evaluated as reliable, i.e., we assume that dependence occurs in this pair.

For C4–I4 (dependence of the car body structural unit on the fault labelled air conditioning), a reliability p value of (0.0499) was obtained for the selected defining factor at the established alpha confidence level.

Since the p value of the pair is <p value set, the dependence can be evaluated as reliable; i.e., we assume that dependence occurs in this pair.

For C5–I9 (dependence of the design unit interior furnishing on the fault labeled power distribution problems), for the selected defining factor at the established alpha confidence level, the reliability p value came out to be (0.0809). Since the p value of the pair > p value determined, the dependence can be evaluated as unreliable; i.e., we do not assume that dependence occurs in this pair.

For C2–(I4 + I6), it is necessary to split the analysis into two planes:

- C2–I4—dependence of the heating and air conditioning design unit on the fault labelled air conditioning—p value = 0.0409;
- C2–I6—dependence of the heating and air conditioning design.

For the chosen defining factor at the established alpha confidence level, the p value for both parameters came out < p value established, and the dependence can be evaluated as reliable; i.e., we assume that dependence occurs in this pair, despite the fact that both failures came out with only a medium level of dependence—i.e., it was not possible to statistically demonstrate dependence between the heating failure and the air conditioning failure. For C3–(I7 + I8 + I9), it is necessary to divide the analysis into three planes:

- C3–I7—dependence of the structural unit electrical equipment and wiring on the fault marked train radio—p value = 0.0750;
- C3–I8—dependence of the electrical equipment and wiring on the fault marked lighting—p value = 0.0231;

- C3–I9—dependence of the design unit electrical equipment and wiring on the fault labelled power distribution problems—p value = 0.0197.

For the chosen defining factor at the established alpha confidence level, the p value for both parameters (d,e) came out < p value established, and the dependence can be evaluated as reliable; i.e., we assume that dependence occurs in this pair. The dependence between failures referred to as power distribution and lighting problems was demonstrated, similar to the above design set.

3.3. There Is a Proven Correlation between Retrofitting and the Reduction in Breakdowns on Passenger Wagons

Due to the gradual modernization of the rolling stock, the following models were selected and compared in terms of their fault rates before and after modernization. Bdteer (18 wagons) and Bmz (26 wagons) were selected. The Bdteer passenger coaches were built and put into service between 1990 and 1992 and have been in service for almost 25 years. As part of the modernization of the passenger wagons, solutions are used to the greatest extent possible to facilitate the unification of the transport operator's fleet. Bmz wagons have been refurbished from Bmeer series wagons. A major shift in technical specification compared to the old wagons occurred in the area of the pipe brake accelerator, which has increased braking power, the boarding doors and front doors, which ensure greater passenger safety, and the information system, which provides external diagnostics of the wagon [17].

In terms of time series, the results were compared, as shown in Table 3.

Table 3. Comparative criteria.

Type of Wagon	Time Series before Modernization	Time Series after Modernization	Numbers of Wagons
Bmz	March 2018–September 2021	October 2021–March 2023	26
Bdteer	January 2020–April 2022	May 2022–March 2023	18

The modernization of both wagon rows was carried out by the same carrier and under the same conditions. During the modernization, several design changes were made, in particular an increase in the design speed, replacement of wiring, modernization of electrical elements such as wagon security, and the modification and replacement of toilets and other structural units. Based on Table 4, the retrofit for the Bdteer series cars has resulted in a reduction in the number of failures on several structural units. The highest benefit was brought by the upgrades to the WC and washroom structural units.

Table 4. Comparison of modernization efficiency on structural elements—Bdteer wagons.

Structural Unit	Proportion of Disorders/1 Month	
	Before Modernization	After Modernization
WC Washroom	7.2143	5.2667
Wagon Cabinet	2.3571	2.2333
Electric Equipment	2.1786	0.9333
Interior Equipment	1.4643	1.0667
Heating/AC	1.1071	0.7667
Others	0.9286	0.3333
Chassis	0.3571	0.2333
Air Brake	0.1786	0.0667
Other Equipment	0.1429	0.0333
Low-Voltage locomotive current	0.0357	0.0333

The failures decreased from an overall average of 7.2 failures per month to 5.27 failures per month. In terms of failures, failures due to malfunctioning toilet equipment have significantly decreased; on the contrary, there have been a number of failures due to blockage of the vacuum toilet and lack of water in the storage tank.

The modernization also led to a significant reduction in faults caused by electrical wiring by almost two per month. Faults caused by insufficient recharging, train radio faults, and lighting problems were eliminated. After the upgrades, the most common fault identified as other faults occurred, which includes mainly broken seals or burnt tubes on lighting bulbs.

The Bmz series upgrades, like the previous series, have been successful. The results shown in Table 5 tell of a reduction in failures on the various passenger car structural units considered.

Table 5. Comparison of modernization efficiency on structural elements—Bmz type wagons.

Structural Unit	Proportion of Disorders/1 Month	
	Before Modernization	After Modernization
WC Washroom	8.5357	4.8667
Wagon Cabinet	3.1429	1.7667
Electric Equipment	2.4643	1.4333
Interior Equipment	2.3571	0.7333
Heating/AC	1.1786	0.5667
Others	0.8929	0.4333
Chassis	0.6429	0.2667
Air Brake	0.3214	0.1333
Other Equipment	0.2143	0.0667
Low-Voltage locomotive current	0.1071	0.033

As with the previous wagon type, there has been a significant reduction in failures on all structural units. In terms of the washbasin and toilet structural unit, there has been a reduction in failures of almost 50%. Failures due to non-functioning toilet equipment and lack of water were reduced. The defects that occurred most frequently after renovation were mainly stickers, interior soiling, and jammed doors.

A significant decrease in defects was also identified on the wagon box structure, where defects due to mechanical damage and soiling were reduced [18].

A decrease in defects was also observed on the structural unit interior equipment after retrofitting—the defects due to problems related to the interior equipment of the wagon were eliminated. After the retrofit, defects were rare, with only problems with wagon cleanliness being identified to a greater extent.

In conclusion, based on the results presented in Tables 4 and 5, the above hypothesis can be accepted, where based on the collected results, it can be stated that the modernization reduced the number of identified failures on the selected structural units of the passenger cars [19,20].

4. Discussion

The modernization of passenger railway wagons is a current topic with the potential to transform the face of railway transportation. It creates a better future for passengers as well as the environment. This discussion focuses on some of the key aspects of modernization and the questions that arise in this significant endeavor [21].

One of the primary objectives of modernization is to enhance passenger comfort. Modernizing the wagon interior can significantly contribute to increasing the number of passengers using rail passenger transport. The modernization of passenger railway wagons has the potential to reshape the way we travel and contribute to a more sustainable future.

Experience in the operation of passenger wagons shows that postponing their modernization and failing to fulfill the repair plan have adverse consequences [22,23]. The process of modernizing the wagons has a linear character; it is supposed to enable effective repairs in accordance with valid repair regulations, i.e., revision repair or intermediate repair. When carrying out general repairs or for specific requirements, equipped pit stations are usually already available in repair shops. Particular emphasis must be placed on the continuous modernization of diagnostic equipment for testing electrical equipment and brake systems to increase the operational reliability of wagons. A specific area in the process of repairing passenger cars is the issue of surface treatment. Particular attention must be paid to its quality. The problems of modernization of wagons and its financing include large-scale rebuilding of interiors, equipment of wagons, bridging of the emergency brake, and installation of fire detectors, central sources or air conditioning [24,25].

Based on the results of the research, it is essential that the wagons currently meet the following conditions. Wagons should be designed for the Central European climate zone, with air-conditioning with cold air streaming from the wagon ceiling and heating and cooling regulation [26]. The next very important condition is windows with tinted athermal windowpanes. After the modernization, to make customers happy, the wagon entrance corridor should be redesigned with modern folding seats and bicycle storage brackets, new easy cleaning floors, an enclosed vacuum toilet system and spacious toilet for wheelchair users, electro-pneumatic plug doors with an integrated system of selective opening to station platform side, electro-pneumatic sliding connecting doors, and sliding interior doors. All doors system should comply with the applicable TSI requirements. An upgraded lighting system and cabling, a new system of sockets for laptops, an upgraded water supply system, and an emergency brake overriding system are needed. There should be a fully air-conditioned interior, 230 V sockets, high performance WiFi for passengers, with all the entrance doors located in the low-floor area. The control system must enable the multiple control of the units. An integral part includes ergonomic seats and a well-arranged information system. An external and internal camera system with remote data transmission is important, and a smart information system is needed [27–29].

Due to rising prices and the energy crisis, repairs and modernization of wagons are taking longer, and the level of quality of rail travel is also decreasing. The level of quality of maintenance provided by suppliers is directly proportional to their technology and innovation. Railway undertakings have set requirements that individual suppliers must meet. The breakdown rate of vehicles is significantly affected by their age, and in this area, the undertakings must try to increase reliability by modernizing the vehicle fleet. Ensuring repairs of railway rolling stock must be approached in a standardized manner and in terms of internal processes with an emphasis on the quality and safety of railway operations. It is necessary to approach individual suppliers in the same way and communicate with them intensively, regardless of whether they are domestic or foreign. Extending repair deadlines naturally has a negative impact on the planning and implementation of traffic services. Contractors must communicate with carriers when dealing with complicated situations and specific requirements in the field of rolling stock repairs. If suppliers do not fulfill their obligations or have a non-standard approach, the solution is to apply contractual fines and sanctions enshrined in individual contracts [30,31].

Based on the previous conclusions of the research, regular and high-quality modernization is part of eliminating the failure rate of wagons. This can only be ensured in the case of a systemic approach aimed at customer orientation and service quality [32].

The methods of the applied statistical analysis have demonstrated that regular maintenance and modernization of railcars are primarily associated with increased passenger safety. Specific technological and structural elements have been identified, and the effectiveness of railcar modernization has reduced the number of malfunctions, thereby minimizing risks and accidents within the operation of passenger rail transportation. The efficiency of modernization is primarily contingent on the level of financial resources provided. Investments in railcar modernization prove to be cost-effective, contributing not only to safer and more efficient operations but also yielding long-term financial benefits for operators of passenger rail transportation. Based on the obtained results, we have identified areas that require further research, particularly in the optimization of specific technological solutions and a deeper understanding of passenger satisfaction with modernized railcars. The results of the statistical correlation confirm that the modernization of passenger railcars brings significant benefits in terms of safety, efficiency, and passenger satisfaction. These findings can serve as a foundation for further innovations in the field of railway transportation and provide support for investments in railcar modernization in the future.

5. Conclusions

The correlation between selected malfunctions in passenger railcars in railway transportation has been thoroughly analyzed and assessed in this study. Based on a wide range of collected data and a detailed examination, mutual relationships between various types of malfunctions and their potential causes have been identified. Statistical correlation has demonstrated a dependence among the selected types of malfunctions, some of which may be caused by similar factors. The theoretical contributions of establishing this correlation enable a better understanding of the interconnection between individual malfunctions, leading to more effective diagnostic procedures and faster fault resolution. Simultaneously, the results of the correlation allow a focus on specific areas where actions for improvement are necessary. Correlations between selected malfunctions in passenger railcars were also identified in the article, with their statistical significance being assessed. The collected data were analyzed using statistical methods, aiming to confirm whether the identified relationships were random or indeed significant. Statistical methods were applied in the research section of the contribution to evaluate the significance of the identified correlations. This approach confirmed that the relationships between malfunctions and the structural components of railway passenger cars were not random but possessed statistical significance. The result of the research also includes a comparative table of railway passenger cars, specifically examining the types of cars Bdteer and Bmz. This table recorded data on the failure rates of the cars for the past month both before and after modernization, considering a selected number of examined vehicles. Following a thorough statistical analysis, a positive impact of the modernization on the operation and failure rates of the cars was indeed observed. For the first series of the Bdteer railway car, the average failure rate was reduced by more than 55% based on the selected indicators. Similarly, for the second type of railway car in the Bmz series, the failure rate decreased by more than 39% under the same statistical parameters. The conclusion of the research emphasizes that understanding and systematically monitoring the structural components of railcars can significantly contribute to preventing malfunctions and improving the overall reliability of the railcar fleet. The findings offer a substantial contribution to more efficient railcar maintenance and can serve as a foundation for future innovations in the construction of passenger railcars, aiming to optimize their performance and ensure safe and reliable railway transportation for passengers.

Author Contributions: Formal analysis, E.N., M.V. and M.M. All authors have read and agreed to the published version of the manuscript.

Funding: This research was funded by [KEGA 010ŽU-4/2023] grant number [010ŽU-4/2023] and the APC was funded by [010ŽU-4/2023].

Data Availability Statement: Data are contained within the article.

Acknowledgments: This paper is a result of the research KEGA 010ŽU-4/2023 Innovative approaches in teaching in the field of transport focused on railway traffic management, with the support of risk and crisis management.

Conflicts of Interest: The authors declare no conflict of interest.

References

1. Dailydka, S. Choosing railway vehicles for carrying passengers. *Transport* **2010**, *25*, 11–16. [CrossRef]
2. Meriç, G. Demiryolu ile Seyahat Eden Yolcuların Haklarına Dair Yönetmelik'in Uygulama Alanı ve Yönetmelik Uyarınca Demiryolu Tren İşletmecisinin Sorumluluğu. *İstanb. Hukuk Mecm.* **2021**, *79*, 837–879. [CrossRef]
3. Ilkjær, L.B.; Lind, T. Passengers' injuries reflected wagon interior at the railway accident in Mundelstrup, Denmark. *Accid. Anal. Prev.* **2001**, *33*, 285–288. [CrossRef] [PubMed]
4. Panahi, E.; Younesian, D. Acoustic performance enhancement in a railway passenger wagon using hybrid ray-tracing and image-source method. *Appl. Acoust.* **2020**, *170*, 107527. [CrossRef]
5. Abood, K.H.A.; Khan, R.A. Railway wagon simulation model to study the influence of vertical secondary suspension stiffness on ride comfort of railway carbody. *Proc. Inst. Mech. Eng. Part C J. Mech. Eng. Sci.* **2011**, *225*, 1349–1359. [CrossRef]
6. Milković, V.; Lisjak, D.; Kolar, D. New reliability-based model of stock optimisation for railroad passenger wagon maintenance. *FME Trans.* **2020**, *48*, 914–921. [CrossRef]
7. Yu, Y.; Zhao, L.; Zhou, C. Modeling and simulation of railway vehicle vertical dynamic behavior by considering the effect of passenger-carbody coupling vibration. *Int. J. Model. Simul. Sci. Comput.* **2019**, *10*, 1950031. [CrossRef]
8. Gallo, F.; Di Febbraro, A.; Giglio, D.; Sacco, N. Planning and optimization of passenger railway services with virtually coupled trains. In Proceedings of the 2021 7th International Conference on Models and Technologies for Intelligent Transportation Systems (MT-ITS), Crete, Greece, 16–17 June 2021; pp. 1–6.
9. Aguayo, P.; Seriani, S.; Delpiano, J.; Farias, G.; Fujiyama, T.; Velastin, S.A. Experimental Method to Estimate the Density of Passengers on Urban Railway Platforms. *Sustainability* **2023**, *15*, 1000. [CrossRef]
10. Craig, M.; Asim, T. Numerical Investigations on the Propagation of Fire in a Railway Carriage. *Energies* **2020**, *13*, 4999. [CrossRef]
11. Liu, T.; Geng, S.; Chen, X.; Krajnovic, S. Numerical analysis on the dynamic airtightness of a railway vehicle passing through tunnels. *Tunn. Undergr. Space Technol.* **2020**, *97*, 103286. [CrossRef]
12. Ulewicz, R.; Nový, F.; Novák, P.; Palček, P. The investigation of the fatigue failure of passenger wagon draw-hook. *Eng. Fail. Anal.* **2019**, *104*, 609–616. [CrossRef]
13. Li, S.; Zhu, X.; Shang, P.; Li, T.; Liu, W. Optimizing a shared freight and passenger high-speed railway system: A multi-commodity flow formulation with Benders decomposition solution approach. *Transp. Res. Part B Methodol.* **2023**, *172*, 1–31. [CrossRef]
14. Leštinský, L.; Zvolenský, P. New methods of noise reduction in railway carriages. *Transp. Res. Procedia* **2019**, *40*, 778–783. [CrossRef]
15. Wei, Z.; Sun, X.; Yang, F.; Ke, Z.; Lu, T.; Zhang, P.; Shen, C. Wagon interior noise-based inspection for rail corrugation on high-speed railway track. *Appl. Acoust.* **2022**, *196*, 108881. [CrossRef]
16. Liu-Henke, X.; Lückel, J.; Jäker, K.-P. An active suspension/tilt system for a mechatronic railway carriage. *Control Eng. Pract.* **2002**, *10*, 991–998. [CrossRef]
17. Krénusz, M.; Kovács, R.; Krémer, M. Estimating the Damage of Railway Wagon Wheels Equipped with Disc Brakes. *Period. Polytech. Mech. Eng.* **2021**, *65*, 286–292. [CrossRef]
18. He, X.; Li, H.; Hu, L.; Wang, H.; Kareem, A. Crosswind aerodynamic characteristics of a stationary interior railway wagon through a long-span truss-girder bridge. *Eng. Struct.* **2020**, *210*, 110350. [CrossRef]
19. Gasparik, J.; Dedik, M.; Cechovic, L.; Blaho, P. Estimation of Transport Potential in Regional Rail Passenger Transport by Using the Innovative Mathematical-Statistical Gravity Approach. *Sustainability* **2020**, *12*, 3821. [CrossRef]
20. Dedík, M.; Gašparík, J.; Záhumenská, Z.; Ľupták, V.; Hřebíček, Z. Proposal of the Measures to Increase the Competitiveness of Rail Freight Transport in the EU. *Naše More* **2018**, *65*, 202–207. [CrossRef]
21. Konečný, V.; Zuzaniak, M.; Brídziková, M.; Jaśkiewicz, M. Regional Differences in the Impact of the COVID-19 Pandemic on the Demand for Bus Transport in the Slovak Republic. *LOGI Sci. J. Transp. Logist.* **2023**, *14*, 146–157. [CrossRef]
22. Pálková, A.; Mašek, J.; Pribula, D. Co-modality as a tool for improvement of public transport. In *Transport Means 2022: Sustainability: Research and Solutions, Proceedings of the 26th International Scientific Conference, Online, 5–7 October 2022*; Kauno Technologijos Universitetas: Kaunas, Lithuania, 2022; Volume 1, pp. 646–650. ISSN 1822-296X.
23. Černá, L.; Pribula, D.; Bulková, Z.; Abramović, B. Draft of Public Rail Passenger Transport During the COVID-19 Pandemic. *LOGI Sci. J. Transp. Logist.* **2023**, *14*, 77–88. [CrossRef]
24. Dömény, I.; Dolinayová, A. Possibilities for Introducing a New Night Train Connection in Middle Europe. *Transp. Tech. Technol.* **2021**, *17*, 6–12. [CrossRef]
25. Buganová, K. Systémy včasného varovania. In Ochrana životních podmínek obyvatelstva, (Early warning systems. In Protection of living conditions of the population). In Proceedings of the 7th International Scientific Conference, Zlín, Czech Republic, 14–17 July 2012; VŠKE: Brno, Czech Republic, 2012; pp. 50–56.
26. Harantová, V.; Mazanec, J.; Štefancová, V.; Mašek, J.; Foltýnová, H.B. Two-Step Cluster Analysis of Passenger Mobility Segmentation during the COVID-19 Pandemic. *Mathematics* **2023**, *11*, 583. [CrossRef]
27. Dolinayova, A.; Domeny, I. Competition on the Railway Market in a Segment of Public Service Obligations in Terms of Effectiveness: Study in V4 Countries. *J. Compet.* **2022**, *14*, 41–58. [CrossRef]
28. Regulation (EC) No 1370/2007 of the European Parliament and of the Council of 23 October 2007 on Public Passenger Transport Services by Rail and by Road and Repealing Council Regulations (EEC) Nos 1191/69 and 1107/70. *Off. J. Eur. Union* **2007**, *L 315*, 1–13.

29. Varcholová, T.; Dubovická, L. *Nový Manažment Rizika*; Wolters Kluwer: Bratislava, Slovakia, 2008; 196p.
30. Hitka, M.; Nad', M.; Langová, N.; Gejdoš, M.; Lizoňová, D.; Sydor, M. Designing chairs for users with high body weight. *BioResources* **2023**, *18*, 5309–5324. [CrossRef]
31. Smerek, L.; Vetráková, M. Difference in Human Resources Development in Various Types of Companies. *Pol. J. Manag. Stud.* **2020**, *21*, 398–411. [CrossRef]
32. Stacho, Z.; Lizbetinova, L.; Stachova, K.; Starecek, A. The Application of Progressive HR Tools in the Environment of Slovak Enterprises. *J. Compet.* **2022**, *14*, 173–190. [CrossRef]

vehicles

Article

An Estimation of the Energy Savings of a Mainline Diesel Locomotive Equipped with an Energy Storage Device

Ievgen Riabov [1,*], Sergey Goolak [2] and Larysa Neduzha [3,*]

[1] Department of Electrical Transport and Locomotive Engineering, National Technical University «Kharkiv Polytechnic Institute», Kyrpychova Street 2, 61002 Kharkiv, Ukraine
[2] Department of Electromechanics and Rolling Stock of Railways, State University of Infrastructure and Technologies, Kyrylivska Street 9, 04071 Kyiv, Ukraine; gulak_so@gsuite.duit.edu.ua
[3] Department of Technical Mechanics, Ukrainian State University of Science and Technologies, Lazaryan 2, 49010 Dnipro, Ukraine
* Correspondence: yevhen.riabov@khpi.edu.ua (I.R.); nlorhen@i.ua (L.N.)

Abstract: The method of improving a two-section mainline diesel locomotive by using energy storage in the traction system is considered. A mathematical model was developed to study the movement of a diesel locomotive based on the recommendations and provisions of the theory of locomotive traction. For this purpose, the movement of a diesel locomotive as part of a train along a given section of a track was studied. It was determined that the use of an energy storage device on a diesel locomotive will allow up to 64% of the energy spent on train traction to accumulate. The use of energy storage in the accumulator during electrodynamic braking ensured a reduction in fuel consumption by about 50%, regardless of the options for equipping the traction system of the diesel locomotive with an energy accumulator. It is established that regardless of the options for equipping the traction system of the diesel locomotive with an energy storage device, the indicators characterizing the degree of use of the diesel engine do not change. These research results can be used in works devoted to the improvement of the control system of energy exchange between the accumulator and traction engines of diesel locomotives.

Keywords: diesel locomotive; energy storage; mathematical modeling; energy saving

1. Introduction

Three international pan-European railway transport corridors—No. 3 (Gdansk–Odesa (Baltic Sea, Black Sea)), No. 5 (Europe–Caucasus–Asia (TRACESA)), and No. 9 (Europe–Asia)—run through the territory of Ukraine. Also, Ukrainian railways through river ports are tangential to pan-European corridor No. 7, which runs along the Danube [1,2]. In 2023, an agreement was reached on the creation of a 1520 mm gauge railway corridor between Ukraine and Lithuania [3]. The railway lines of the active transport corridors, as well as the main highways, are electrified. This ensures a low cost of transportation and the potential for further development of transit transportation [4]. At the same time, parts of the railway lines that transport goods to stations located on the lines of transport corridors, ports, logistics terminals, etc., are served by diesel locomotives. The insufficient volumes of these vehicles make it impractical to electrify such areas, even under conditions of constant growth in the cost of fuel and lubricants. Therefore, it is considered that, for such lines, it is necessary to use modern or modernized diesel locomotives; as a result, a reduction in operating costs and the elimination of the negative impact on the environment will be achieved.

To update the fleet of mainline freight diesel locomotives, TE33A "Tryzub" diesel locomotives were purchased, which are a version of the Evolution ES44ACi diesel locomotive adapted for 1520 mm gauge [5,6]. The advantages of the new diesel locomotive are reduced consumption of fuel and lubricants, increased service intervals, lower harmful emissions, improved working conditions of the locomotive crew, etc. Also, the new diesel locomotive has significantly higher traction characteristics compared to existing diesel locomotives, which makes it possible to increase the efficiency of the transportation process.

Another option for renewal is the modernization of diesel locomotives, which is widely used to improve the locomotive fleet. Its advantage is a lower cost compared to the purchase of a new locomotive. The diesel locomotive fleet of state railways and private carriers in the Baltic States and Eastern Europe has been updated through modernization. In 2011, the domestic railway operator JSC "Ukrzaliznytsia" modernized M62 diesel locomotives using General Motors equipment [7]. The PrJSC Poltava Diesel Locomotive Repair Plant modernized 2TE10M diesel locomotives using the SuperSkid power module from GE [8]. Modernization ensured a reduction in fuel consumption, reduced oil consumption, etc. However, due to the preservation of the original traction electric motors, the traction properties of diesel locomotives have hardly improved.

The method of updating the fleet of locomotives is to reduce operating costs, a significant part of which is the cost of fuel and lubricants. The key here is to ensure the use of modern diesel engines, which use less fuel compared to older models. The use of traction electric drives and auxiliary electric drives, which have increased energy efficiency, also contribute to reducing fuel consumption. A reserve for further reductions in fuel consumption is the accumulation of energy during electrodynamic braking and its use in traction modes. This technology is being researched by many research centers and companies [9].

In existing studies, two approaches to the use of energy storage devices are proposed [9–11]. The first is that the energy storage is integrated into the diesel generator set, and they can work on a common load [12–14]. The second involves energy storage that is placed on a separate section and interacts only with the traction drive of this section [15–17].

The results of the study in accordance with the first option of adding an energy storage device to the traction system can be found in [10]. In this paper, through modeling, it is established that the use of an energy storage device leads to a reduction in fuel consumption by up to 16.5%, and harmful diesel emissions are also reduced. In [11], it is shown that the potential energy saving during the movement of the studied train is 23.87%. Taking into account the construction of the optimal cost criterion of the life cycle of the drive in LTO elements, the savings will be 15.67%.

Research on the application of energy storage is applied to many types of rail transport [18–20]. In [21], a shunting locomotive is studied when performing removal work. It is shown that when using an energy storage system in the traction system, the reduction in fuel consumption is 20.57% without the use of recuperation and 25.64% in the case of using energy storage. In [22], the results of the operation of the Ki-Ha E200 rail bus are given, the use of a storage device on which saved 10% of fuel. Ref. [23] provides data on the reduction in energy consumption by a passenger diesel locomotive for various operating conditions. The reduction in energy consumption is 47.52% of the total energy consumption.

An alternative option is the use of a separate booster section in which the energy storage is located [17,24,25]. Operational studies of the FLXdrive battery locomotive are being conducted in the USA, the use of which as part of a multi-section diesel locomotive made it possible to save 11% of fuel [15]. The research results on a locomotive with a booster section equipped with an energy store are given in [17,26]. According to the simulation results, it was established that fuel consumption by locomotives from energy storage is reduced by 12.40% when using inertial storage and 20.65% when using batteries. In [25], it was found that the use of a booster section with an energy store allows fuel consumption to be reduced by 22–30%.

Thus, the results of theoretical and experimental studies show the possibility of eliminating the consumption of fuel and energy resources when using energy accumulators on mainline freight locomotives. The limitation of the obtained results is that they were obtained for a specific type of diesel locomotive under specific operating conditions. This makes it impossible to extend them to other conditions, and therefore, there is a need for new studies on specific types of locomotive, train, and route. To determine the effectiveness of the use of energy storage on rolling stock, indicators characterizing the costs of operating a serial rolling stock without energy storage and a hypothetical modernized rolling stock with energy storage are compared [10,23]. As a rule, fuel consumption is taken as an indicator. In more advanced calculations, economic indicators and combinations of indicators that take into account various aspects of locomotive operation are used as efficiency criteria [11,16]. Considering the above, it is necessary to carry out research to determine the effectiveness of energy storage on domestic rolling stock.

The aim of this research is to assess the potential of energy saving when using energy storage devices on mainline diesel locomotives on the railways of Ukraine.

The following tasks were performed as part of the assigned work:

- A mathematical model of the movement of the main diesel locomotive is developed in Section 2;
- The movement of a diesel locomotive as part of a train is modeled for the following options in Section 3: movement with a serial diesel locomotive; movement with a diesel locomotive as part of one diesel and booster section with an energy accumulator (energy exchange is carried out only with the traction system of the booster section); movement with a diesel locomotive as part of one diesel and booster section with an energy accumulator (energy exchange is carried out with the traction systems of the diesel and booster sections);
- The obtained results are discussed in Section 4;
- Conclusions are drawn in Section 5.

2. Development of a Mathematical Model of Train Movement

The most common locomotives on Ukrainian railways are the 2TE116 freight locomotives manufactured by PJSC "Luhanskteplovoz" (Luhansk, Ukraine) (Figure 1). The basic form is a two-section diesel locomotive. The design of the locomotive is described in [27–29].

Figure 1. General appearance of the 2TE116 diesel locomotive (photo [30]). Non-English word means an element of the locomotive livery, which contains the name of the company in Ukrainian.

This study aimed to determine the movement parameters of the mainline diesel locomotive as part of a train journey carried out in the section between the Kharkiv-Sortuvalnyi and Sumy stations in both directions. The length of the section is 201 km; the profile of the route is shown in Figure 2.

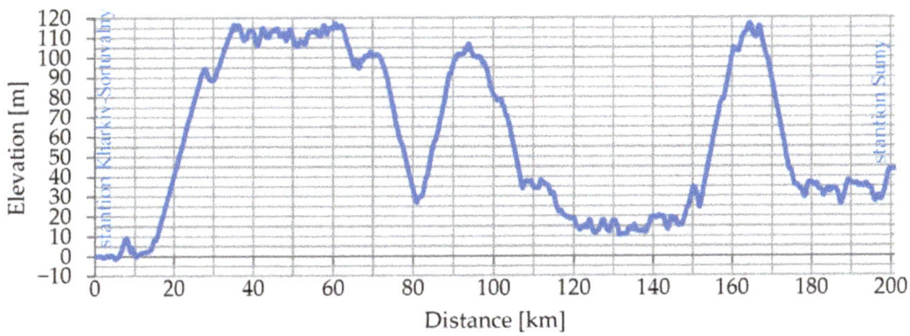

Figure 2. Path profile (created by the authors).

An analysis of works devoted to the research of new technologies on rolling stock shows that mathematical modeling is used in the initial stages [16,19]. Modeling is important in creating modern equipment for vehicles [31,32] and rail transport systems in general [32–34]. Mathematical modeling is widely used, in particular, for the study of rolling stock [35–37], traction power supply systems [38–40], infrastructure objects [41–43], traffic safety assessment [44,45], transportation process management [46], etc. The priority is mathematical modeling that does not require the production of physical prototypes of equipment and systems.

Mathematical modeling is used in research devoted to the application of energy storage on railway rolling stock [16,21]. This makes it possible to assess the expediency and efficiency of energy storage applications. For this purpose, performance indicators of a serial locomotive without energy storage and a hypothetical locomotive equipped with energy storage are compared [16,19,25].

The mathematical modeling of train movement involves determining movement parameters: duration, speed, the consumption of energy resources, etc. A mathematical model was developed to study the movement of a mainline locomotive based on the recommendations and provisions of the theory of locomotive traction [47]. It is accepted that the train was viewed as a chain of rigid bodies connected rigidly [48].

The mathematical model of train movement is described by the following system of equations [48,49]:

$$\begin{cases} \frac{dv}{dt} = \frac{\xi}{\rho}(f_L - (w_L + w_W) - b); \\ \frac{dS}{dt} = v, \end{cases} \tag{1}$$

where ξ—coefficient taking into account the units of measurement;

v—train speed;
t—time;
S—path;
ρ—coefficient that takes into account the rotation of the parts of the locomotive subframe;
f_L—specific traction force of the locomotive in traction or electrodynamic braking mode;
w_L—specific resistance force of the locomotive movement;
w_W—specific force of resistance to movement of wagons;
b—specific braking force of pneumatic brakes.

The specific tangential force of the locomotive in the mode of traction or electrodynamic braking was determined in accordance with the expression

$$f_L = \frac{F_L}{\sum\limits_{k=1}^{s} M_{Lk} + \sum\limits_{j=1}^{n} M_{Wj}} \tag{2}$$

where F_L—traction force of the locomotive in traction or electrodynamic braking mode;

M_{Lk}—mass of the locomotive section;
s—the number of locomotive sections;
M_{Wj}—mass of the wagon;
n—the number of wagons.

The specific resistance to movement of the locomotive and wagons was determined in accordance with the expression

$$w = w_L + w_w + w_i + w_r + w_s, \tag{3}$$

where w_L—main specific resistance of movement;

w_i—additional specific resistance to movement from a slope;
w_r—additional specific resistance of movement from movement along a curve;
w_s—additional specific resistance during train movement.

The main specific resistance during locomotive movement was determined as follows [47]:

$$w_L = \begin{cases} 17.2 + 0.044 \cdot v + 0.0026 \cdot v^2, \ F_L > 0; \\ 20.5 + 0.18 \cdot v + 0.0034 \cdot v^2, \ F_L = 0. \end{cases} \tag{4}$$

The first equation of the system corresponds to the traction mode, and the second to the coasting and braking modes.

The main specific resistance of four-axle wagons on semi-wagons was determined as [48]

$$w_W = \left(0.7 + \frac{3 + 0.09 \cdot v + 0.002 \cdot v^2}{q_0}\right) \cdot g, \tag{5}$$

where q_o—axle load;

g—acceleration of free fall.

With the calculated weight of the loaded wagon of 84 tons, the axle load is 21 tons.

In the case of the movement of an empty wagon (or with an axle load of less than 6 tons), the main specific resistance was calculated in accordance with the following expression [48]:

$$w_{Wx} = 3 + 0.1 \cdot v + 0.0025 \cdot v^2. \tag{6}$$

Additional resistance when moving on a slope was determined as

$$w_i = i \cdot g, \tag{7}$$

where i—axle load.

Let us determine the resistance from movement along a curved section of the track in accordance with the expression [48]

$$w_r = \frac{700}{R} \cdot g, \tag{8}$$

where R—radius of the curve.

Additional resistance during displacement was determined in accordance with the expression [48]

$$w_s = \frac{28}{q + 7} \cdot g, \tag{9}$$

where q—load on the axis of the rolling stock unit.

The control system, which sets the force of traction and braking, functions in such a way as to maintain the permissible speed of movement on the site. Stop braking is carried out at the end of the movement on a section that is 1500 m long.

The energy that is actually produced by the diesel during movement in traction modes is

$$G = \sum_{i=1}^{N} g_i \cdot \tau_i + g_0 \cdot \tau_0, \tag{10}$$

where i—position number of the driver controller in traction mode;

g_i—specific fuel consumption at the i-th position of the driver's controller;
τ_i—total duration of diesel operation at the i-th position of the driver's controller;
g_0—specific fuel consumption in idle mode;
τ_0—total duration of diesel engine operation in idling mode;
N—total number of driver controller positions.

For additional evaluation of the use of the power plant, the following indicators are used.

The coefficient of use of the installed diesel power, which characterizes the completeness of the consumption of the installed (nominal) power of the diesel generator set when moving under load, is defined as [50,51]

$$K_P = \frac{E_F}{E_D}, \tag{11}$$

where E_F—energy actually produced by the diesel engine during movement in traction mode;

E_D—theoretical value of the energy that a diesel engine could produce when operating at rated power during movement.

The energy actually produced by the diesel engine during movement in traction mode was determined as

$$E_F = \sum_{i=1}^{N} P_i \cdot \tau_i, \tag{12}$$

where P_i—power of the diesel engine at the i-th position of the driver's controller.

The theoretical energy value is calculated according to the expression

$$E_D = P_N \cdot \sum_{i=1}^{N} \tau_i, \tag{13}$$

where P_N—nominal power of the diesel engine.

The load factor of the diesel engine over time [50] was determined as

$$K_T = \frac{T_1}{T_1 + T_2 + T_3}, \tag{14}$$

where T_1—total duration of work under load in traction mode;

T_2—total duration of coasting mode during movement along the section;
T_3—total duration of braking mode, including electrodynamic braking.

The theoretical coefficient of energy accumulation was determined by the expression

$$k_{RT} = \frac{A_B}{A_T} \cdot \eta_R \cdot (1 - \gamma_{AUX}) \cdot (1 - \gamma_B), \tag{15}$$

where A_B—work "on the wheel" of braking forces;

A_T—work "on the wheel" of traction forces;
η_R—average value of the efficiency of the traction system and the energy storage system during accumulation;
γ_{AUX}—share of energy consumed by auxiliary systems in electrodynamic braking mode;

γ_B—share of energy that is utilized by other braking systems during braking.

The coefficients η_R, γ_{AUX}, and γ_B can be reliably determined based on the results of calculations of parameters of motion processes. Let us consider it expedient to accept the following values for estimation calculations:

- The average value of the efficiency of the traction system and the energy storage system during accumulation is 0.9;
- The share of energy consumed by auxiliary systems in electrodynamic braking mode: $\gamma_{AUX} = 0.05$;
- The share of energy that, during braking, is utilized by other braking systems: $\gamma_B = 0.06$.

Then, Expression (15) takes the form

$$k_{RT} \approx 0.8 \cdot \frac{A_B}{A_T}. \tag{16}$$

The actual accumulation coefficient was determined in accordance with the expression

$$k_{RF} = \frac{A_E}{A_T}, \tag{17}$$

where A_E—energy that was actually accumulated in the accumulator.

3. Results of Train Motion Simulation

3.1. Modeling the Movement of a Train with a Serial Diesel Locomotive (Option 1)

The electric locomotive uses alternating current–direct current power transmission. The diesel generator has a traction synchronous generator in its composition, which feeds an uncontrolled rectifier. DC traction motors with series excitation are powered from the rectifier. The regulation of electric motors is carried out by changing the voltage on their armatures, for which the excitation current of the traction generator is changed. In addition, shunting of the excitation windings of traction electric motors is used, for which field weakening resistors and appropriate contact equipment are used. A simplified structural diagram of the traction system of a serial diesel locomotive is shown in Figure 3. The traction systems of sections 1 and 2 are identical.

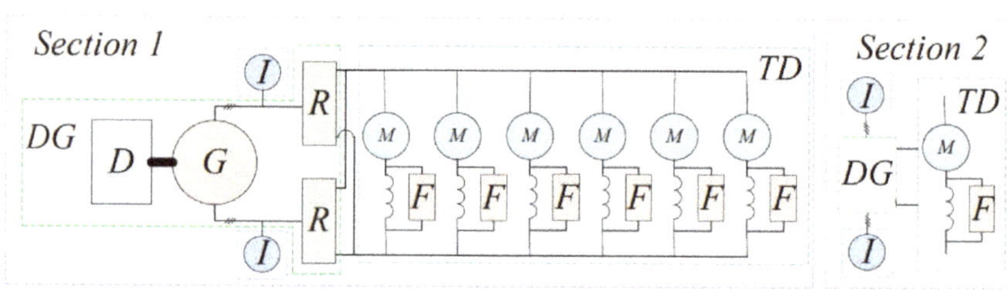

Figure 3. Simplified structural diagram of the traction system of the 2TE116 diesel locomotive.

In Figure 3, the following designations are used: *DG*—diesel genset with non-controlled rectifier, *TD*—traction electric drive, *D*—diesel engine, *G*—traction synchronous generator, *R*—non-controlled rectifier, *M*—DC traction electric motors, *F*—field weakening resistors and their switching devices, and *I*—auxiliary induction electric motors of equipment-cooling motor fans.

Figure 4 shows the traction characteristics of the 2TE116 serial diesel locomotive in the composition of two identical sections.

Table 1 shows fuel consumption on hyperbolic sections of traction characteristics.

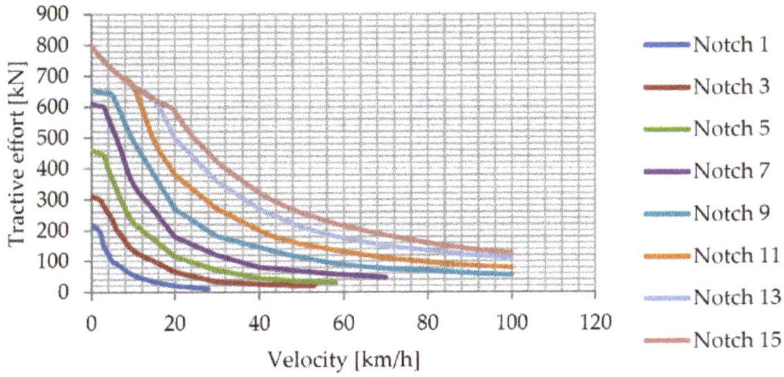

Figure 4. Traction characteristics of the 2TE116 diesel locomotive at odd positions of the driver's controller.

Table 1. Characteristics of fuel consumption by diesel locomotive 2TE116 in the composition of two sections.

Driver controller position	0	1	3	5	7	9	11	13	15
Power of diesel engines, kW	2 × 33	2 × 180	2 × 415	2 × 670	2 × 990	2 × 1320	2 × 1660	2 × 2000	2 × 2250
Fuel consumption by two diesel engines, kg/min	0.66	1.6	3.3	4.8	6.8	9.0	10.9	13.5	17.0

Calculations were made for the permissible speed of 60 km/h, which corresponds to the actual speeds of freight trains on the section. For the calculation, a train weighing 4056 tons with forty-five loaded wagons (this corresponds to the weight standard for a 2TE116 diesel locomotive) and 1311 tons with forty-five empty wagons was accepted. The mass of the diesel locomotive is 276 tons. The simulation was carried out while moving in both directions. The results of calculations are given in Table 2.

Table 2. Results of the calculation of traffic parameters on the Kharkiv-Sortuvalnyi–Sumy section.

Parameter	**Movement Direction**			
	Kharkiv-Sortuvalnyi–Sumy		**Sumy–Kharkiv-Sortuvalnyi**	
Type of Train	**Loaded Train**	**Empty Train**	**Loaded Train**	**Empty Train**
Duration of movement, s	13,264	12,483	13,163	12,448
Fuel consumption, kg	1726	837	1611	780
Work "on the wheel" in traction mode, kWh	5605.3	2081.9	5173.9	1940.4
Work "on the wheel" of braking forces, kWh	3830.6	1071.5	4206.1	1190.0
Theoretical accumulation coefficient	0.544	0.423	0.648	0.488
The coefficient of use of the installed diesel power	0.881	0.524	0.876	0.521
Diesel engine load factor over time	0.501	0.434	0.482	0.411

From Figures 5 and 6, it can be seen that the movement is carried out at a permissible speed. For loaded wagons, there is a decrease in speed when moving on climbs; the tangential power mostly corresponds to the nominal power of the diesel locomotive. When moving with empty wagons, the tangential power takes intermediate values according to the locomotive's traction characteristics. For the reverse direction—from the Sumy station to the Kharkiv-Sortuvalnyi station—the speed and traction force dependences are similar to those shown in Figures 5 and 6.

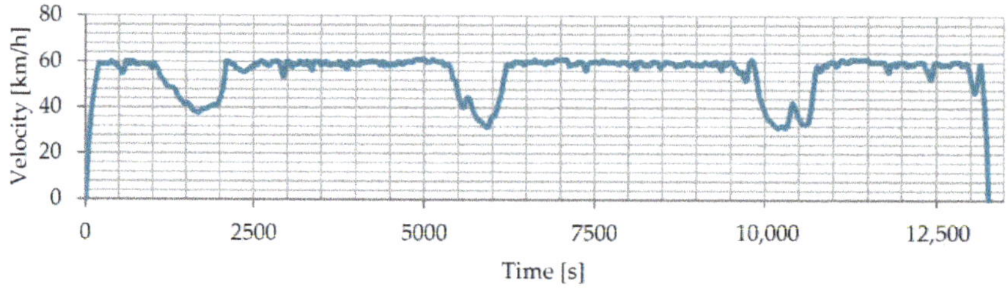

(a)

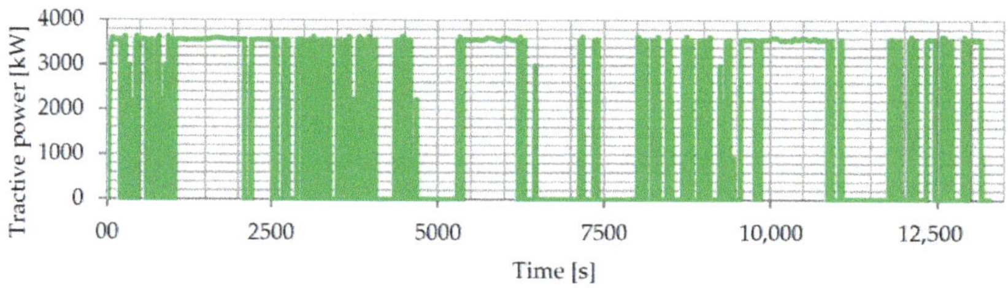

(b)

Figure 5. Time dependence of the speed (**a**) and tangential power of the diesel locomotive (**b**) during the movement of a train with loaded wagons from the Kharkiv-Sortuvalnyi station to the Sumy station.

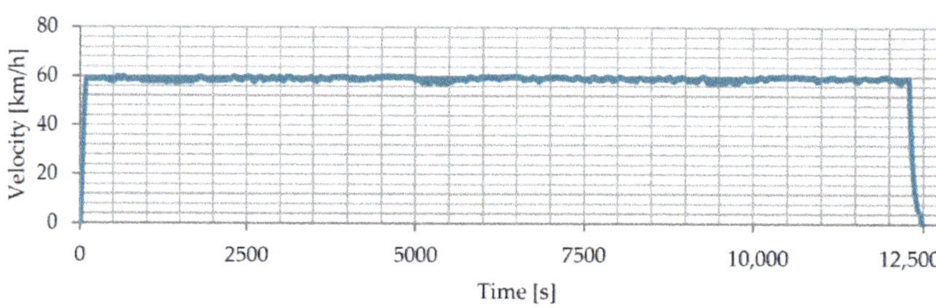

(a)

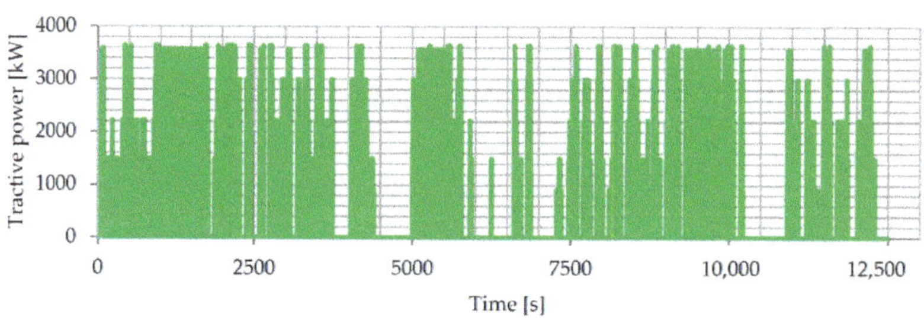

(b)

Figure 6. Time dependence of the speed (**a**) and tangential power of the diesel locomotive (**b**) during the movement of a train with empty wagons from the Kharkiv-Sortuvalnyi station to the Sumy station.

The data in Table 2 shows that, for movement with loaded wagons, the coefficient of use of the installed diesel power is 0.87–0.88, and the loading coefficient of the diesel engine over time is 0.48–0.5. When moving with empty wagons, the utilization factor of the installed diesel power is 0.52, and the load factor of the diesel engine over time is 0.41–0.43. Larger values of the coefficient of use of the installed capacity of the diesel engine correspond to more difficult driving conditions with loaded wagons. However, the diesel load factor changes insignificantly over time when the mass of the train changes.

The theoretical coefficient of energy accumulation is 0.426–0.648. That is, 42.6% to 64.8% of the energy spent in traction mode can be stored in the on-board energy storage system and can be reused in traction modes. This will ensure a reduction in fuel consumption and increase the service life of the diesel engine. To implement this, the traction system of the diesel locomotive must include an energy storage system.

3.2. Modeling the Movement of a Train with a Diesel Locomotive as Part of Diesel Sections and a Booster Section with Energy Storage (Option 2)

Studying the movement of a train with a serial diesel locomotive showed the presence of a significant potential for accumulating energy in the on-board storage device. The option was studied when one of the sections of the diesel locomotive was converted and a storage device was installed on it. For this purpose, it is necessary to use semiconductor converters for powering and controlling electric motors, which, in this case, will work with current pulsations [52,53]. A possible version of the construction of the traction system of the locomotive is shown in Figure 7 (the diesel section does not change; its traction system is shown in Figure 3).

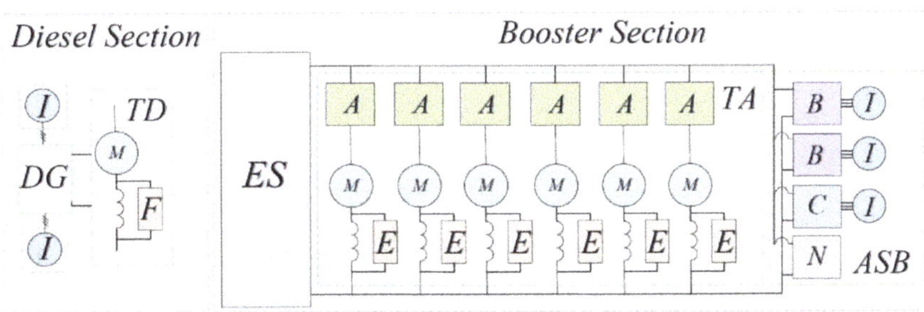

Figure 7. Structural diagram of the traction system of diesel locomotive with isolated booster section.

In Figure 7, the following designations are used: *TA*—traction electric drive with semiconductor converter, *ES*—energy accumulator, *A*—semiconductor converter of the armature circuit, *E*—semiconductor excitation regulator, *ASB*—auxiliary system of the booster section, *B*—semiconductor power converter of cooling motor fans of traction electric motors, *C*—semiconductor power converter of the compressor unit, *N*—semiconductor power converter of additional systems of the booster section, and *I*—auxiliary induction electric motors. Other designations are shown in Figure 3.

For modeling purposes, it is assumed that the booster section with energy storage has traction and braking characteristics identical to the characteristics of the diesel section. The traction characteristics of the booster section are shown in Figure 8.

Since, in the studied version of the locomotive, the diesel generator set remains only on the diesel section, the fuel consumption of the locomotive will decrease (Table 3).

Table 3. Fuel consumption by diesel section.

Driver controller position	0	1	3	5	7	9	11	13	15
Diesel power, kW	33	80	415	670	990	1320	1660	2000	2250
Fuel consumption, kg/min	0.33	0.8	1.7	2.4	3.4	4.5	5.9	6.8	8.5

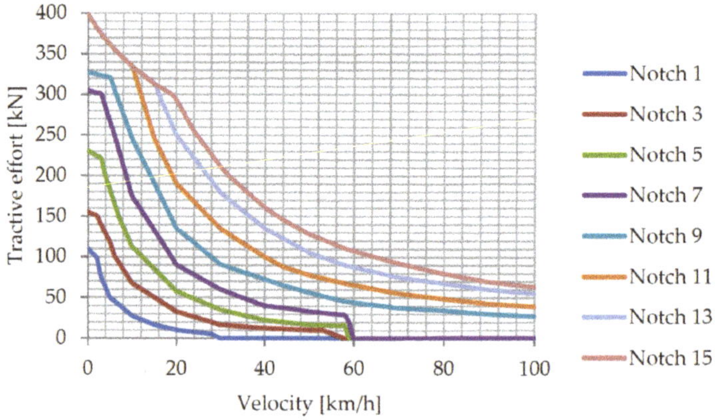

Figure 8. Traction characteristics of the booster section of the diesel locomotive.

The limiting braking characteristic of one section of the diesel locomotive is shown in Figure 9.

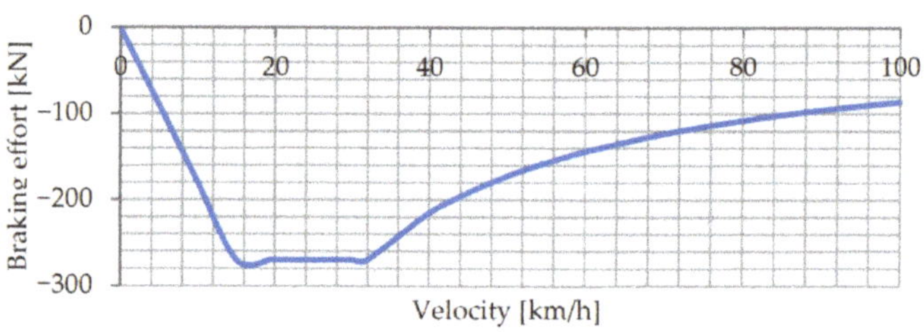

Figure 9. Limiting braking characteristic of one section of a diesel locomotive.

With electrodynamic braking, the operation of both sections is possible at any point within the braking area [54]. On the accelerated section, this is ensured by appropriate control of the traction converters. To implement electrodynamic braking, braking resistors must be installed on the diesel section. If the section is equipped with an automatic rheostat brake control system, it is also possible to operate in any mode corresponding to the points "inside" the area that are limited by the limiting braking characteristic. The scheme of the traction wheels in the braking mode corresponds to the traditional schemes that are used on diesel locomotives with rheostat braking and are not shown here. It should be noted that for cooling the traction electric motors in the rheostat braking mode, an increase in the engine speed is necessary. There are no data on fuel consumption in such modes, so in the simulation, it is assumed that the fuel consumption corresponds to the mode of operation at the zero position of the driver's controller (although the real fuel consumption will be greater).

Table 4 shows the results of calculations during the movement of a diesel locomotive, which consists of serial diesel and booster sections with energy storage. The simulation is carried out under the same conditions as in option 1.

The calculation of energy exchange processes was carried out as follows [55,56].

The power, which is exchanged by the storage unit with the traction electric drive, was determined as

$$P_{ES} = \begin{cases} \frac{P_{kB}}{\eta_{TD}} + P_{auxB}, \ P_{kB} \geq 0; \\ -(|P_{kB}| \cdot \eta_{TD} - P_{auxB}), \ P_{kB} < 0, \end{cases} \quad (18)$$

where P_{kB}—tangential power of the booster section;

η_{TD}—total efficiency of traction electric drive (accepted efficiency equal to 0.9 for traction mode and electrodynamic braking mode);
P_{auxB}—power of auxiliary systems of the booster section (taken equal to 150 kW).

Table 4. Results of calculation of train movement parameters.

Parameter	Movement Direction			
	Kharkiv-Sortuvalnyi–Sumy		Sumy–Kharkiv-Sortuvalnyi	
Type of Train	**Loaded Train**	**Empty Train**	**Loaded Train**	**Empty Train**
Duration of movement, s	13,345	12,524	13,212	12,461
Fuel consumption, kg	865	422	806	398
Work "on the wheel" in traction mode, kWh	5603.1	2084.9	5162.5	1938.5
Work "on the wheel" of braking forces in EDB mode, kWh	3831.2	1070.1	4216.7	1191.0
Theoretical accumulation coefficient	0.544	0.423	0.648	0.488
Coefficient of use of the installed diesel power	0.881	0.524	0.876	0.521
Diesel engine load factor over time	0.501	0.434	0.482	0.411
Energy accumulated in the accumulator at EDB, kWh	1336.4	473.4	1550.7	520.0
Actual accumulation rate	0.238	0.227	0.301	0.268
Energy consumed from ES in traction modes, kWh	3554.4	1642.1	3288.2	1555.1
Energy that must be stored in the ES before movement, kWh	2110	1115	2001	1013
Minimum working energy capacity, kWh	2100	1115	2001	1013

The energy change in the accumulator was determined by the expression

$$\Delta E_{ES} = \begin{cases} -P_{ES} \cdot \eta_{ES} \cdot \Delta t, \ P_{ES} > 0; \\ \frac{|P_{ES}| \cdot \Delta t}{\eta_{ES}}, \ P_{ES} < 0, \end{cases} \tag{19}$$

where η_{ES}—energy storage efficiency (in charging and discharging modes, this is equal to 0.9); Δt—duration of the charging (discharging) process.

Table 4 shows that the parameters correspond to the traffic parameters with a serial diesel locomotive. Minor differences can be explained by modeling errors. Table 4 shows that the actual coefficients of energy accumulation are approximately half as much as the theoretical values. This is explained by the fact that energy is accumulated only from the traction electric drive of the booster section. The diesel section does not provide energy for the accumulator. However, fuel consumption was reduced by approximately half. That is, the use of an energy storage device made it possible to halve the consumption of fuel, which is consumed directly during the movement of a diesel locomotive with a train along the section.

Figures 10 and 11 show the time dependences of power and energy changes in the accumulator. The motion curves (speed, tangential power) are identical to the curves obtained during the movement of a train with a basic diesel locomotive (Figures 5 and 6). Table 4 shows the results of calculating the movement parameters for the studied option.

The analysis of the dependencies in Figures 10 and 11 shows that intensive energy consumption occurs when moving uphill, and accumulation occurs when moving downhill. The maximum power of the energy storage in the traction mode of the booster section is 2156 kW, and in the energy storage mode with electrodynamic braking, this value is 2337 kW. To ensure the operation of the storage device, it must have a certain energy charge before starting [56,57]. At the same time, there is an almost monotonous change in energy in the accumulator, which allows us to estimate its minimum working energy capacity equal to the initial energy before movement.

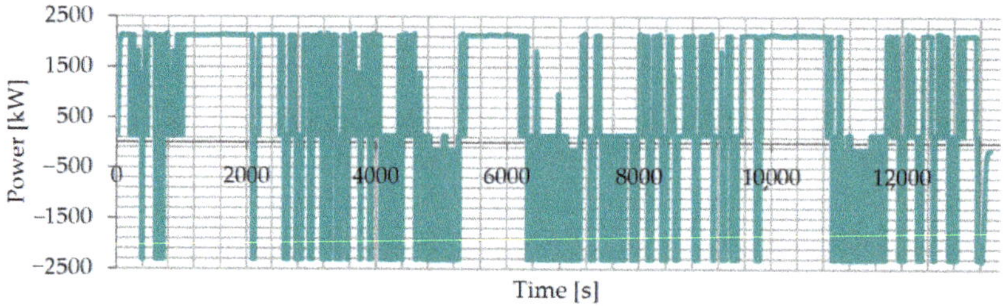

(a)

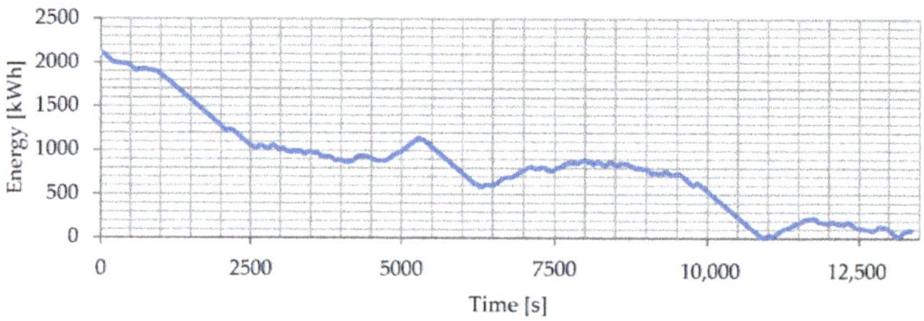

(b)

Figure 10. Time dependences of power (**a**) and energy change (**b**) of the accumulator when moving from Kharkiv-Sortuvalnyi station to Sumy station with loaded wagons.

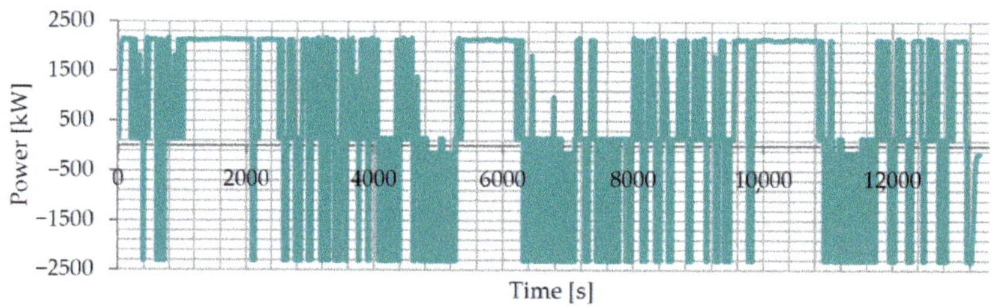

(a)

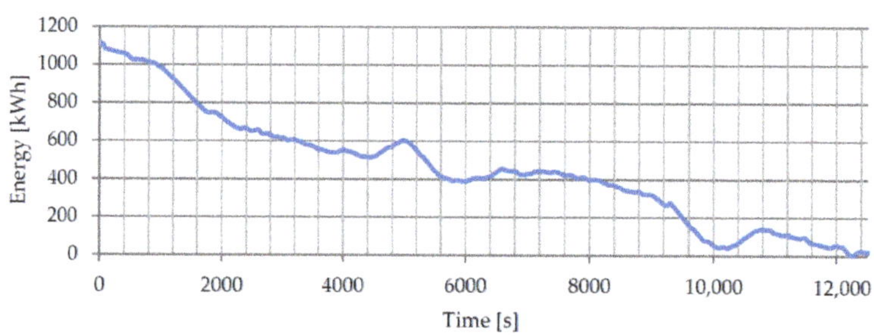

(b)

Figure 11. Time dependences of power (**a**) and energy change (**b**) of the accumulator when moving from the Kharkiv-Sortuvanny station to the Sumy station with empty wagons.

Thus, the use of energy storage on the booster section reduced fuel consumption during movement by half. As a result of the fact that accumulation is only carried out from the booster section, about half of the available energy is stored. The performance indicators of the diesel engine do not change.

3.3. Modeling the Movement of a Train with a Diesel Locomotive as Part of a Diesel Section and a Booster Section with an Energy Storage Unit with Combined Traction Systems (Option 3)

As can be seen from the analysis of the previous simulation results, energy accumulation from only one section does not exhaust the entire energy saving potential. Therefore, the case was considered when, in the mode of electrodynamic braking, energy accumulation can occur both from the booster section and from the diesel section. At the same time, the energy storage is located in the booster section. In this case, the traction system of the diesel section will have a structure similar to the structure of the booster section (Figure 12).

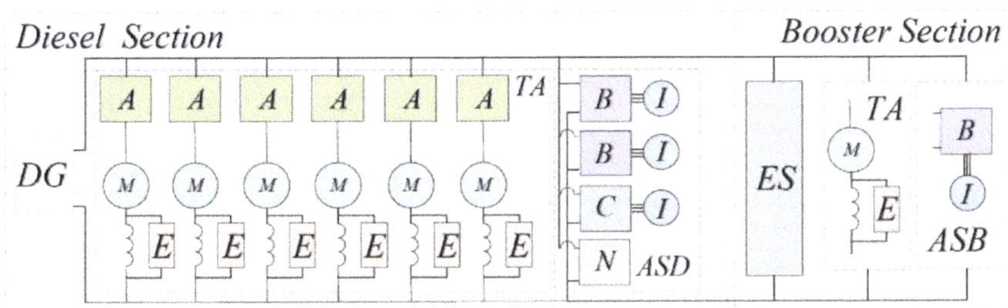

Figure 12. Structural diagram of the traction system of the diesel section for the case of energy accumulation both from the booster section and from the diesel section during electrodynamic braking.

In Figure 12, the following designation is used: *ASD*—auxiliary system of the booster section. Other designations are shown in Figures 3 and 7.

To provide energy return from the electric motors of the diesel section to the accumulator, which is located in the booster section, the power circuits of the diesel and booster sections are connected into a single link. At the same time, in the traction mode, only the traction electric motors of the booster section receive power from the energy storage.

The traction characteristics of the sections shown in Figure 8 and the limiting braking characteristics shown in Figure 9 were used in the simulation.

The calculation of energy exchange processes in the energy storage is carried out as follows.

The power that is exchanged between the storage unit and the traction electric drive is

$$P'_{ES} = \begin{cases} \frac{P_{kB}}{\eta_{TD}} + P_{auxB},\ P_{kB} \geq 0; \\ -((|P_{kD}| + |P_{kB}|) \cdot \eta_{TD} - (P_{auxD} + P_{auxB})),\ P_k < 0, \end{cases} \quad (20)$$

where P_{kD}—tangential power of the diesel section;

P_{auxD}—the power of auxiliary systems of the diesel section (taken to be equal to 200 kW);
P_k—tangential power of the diesel locomotive.

The total efficiency of the traction electric drive, η_{TD}, is taken to be equal to 0.90 for the traction mode and the electrodynamic braking mode for both the booster and diesel sections.

Positive values of storage capacity correspond to discharging, with negative values to charging.

The change in energy in the accumulator is determined by the expression

$$\Delta E'_{ES} = \begin{cases} -P'_{ES} \cdot \eta_{ES} \cdot \Delta t, \; P'_{ES} > 0; \\ \frac{|P'_{ES}| \cdot \Delta t}{\eta_{ES}}, \; P'_{ES} < 0. \end{cases} \quad (21)$$

As in the previous cases, the simulation was performed for the movement of the train with loaded and empty wagons in both directions. The calculation results are given in Table 5. Figures 13 and 14 show the time dependences of power and storage energy changes for the studied variant.

The analysis of movement parameters according to the data in Table 5 shows that they correspond to the traffic parameters with a serial diesel locomotive.

Table 5 shows that the actual coefficients of energy accumulation are practically equal to the theoretical value (the excess of the actual coefficient over the theoretical one is explained by the approximate calculation of its value). This is explained by the fact that energy is accumulated both from the traction electric drive of the booster section and from the diesel section. Fuel consumption by the diesel section is reduced by about half.

From Figures 10 and 11, it can be seen that intensive energy consumption occurs when moving uphill, and accumulation occurs when moving downhill. The maximum power of the energy storage in the traction mode of the booster section is 2156 kW, and in the energy storage mode with electrodynamic braking, this value is 4674 kW. To ensure the operation of the drive, it must have a certain charge of energy before starting. The analysis of the dependences of the energy change in the accumulator shows that it is characterized by a significant discharge of the energy accumulator as well as charging. At the same time, energy accumulates in the accumulator at the end of the movement. The minimum working energy capacity can be estimated at the level of the initial energy that must be in storage before starting the movement. It should be noted that the values of the working capacity of the drive for option 3 are significantly lower than those for the drive used according to option 2.

Table 5. The change in energy in the accumulator is determined by the expression.

Parameter	Movement Direction			
	Kharkiv-Sortuvalnyi–Sumy		Sumy–Kharkiv-Sortuvalnyi	
Type of Train	**Loaded Train**	**Empty Train**	**Loaded Train**	**Empty Train**
Duration of movement, s	13,345	12,524	13,212	12,460
Fuel consumption, kg	865	423	806	398
Work "on the wheel" in traction mode, kWh	5603.1	2084.9	5162.5	1938.5
Work "on the wheel" of braking forces in EDB mode, kWh	3823.2	1072.5	4185.7	1236.0
Theoretical accumulation coefficient	0.546	0.411	0.649	0.509
Coefficient of use of the installed diesel power	0.881	0.524	0.876	0.521
Diesel engine load factor over time	0.501	0.434	0.482	0.411
Energy accumulated in the accumulator at EDB, kWh	2672.8	946.7	3288.2	1040.1
Actual accumulation rate	0.477	0.454	0.637	0.536
Energy consumed from ES in traction modes, kWh	3550.0	1642.1	3101.4	1555.1
Energy that must be stored in the ES before movement, kWh	1040	710	890	605
Minimum working energy capacity, kWh	1040	710	890	605

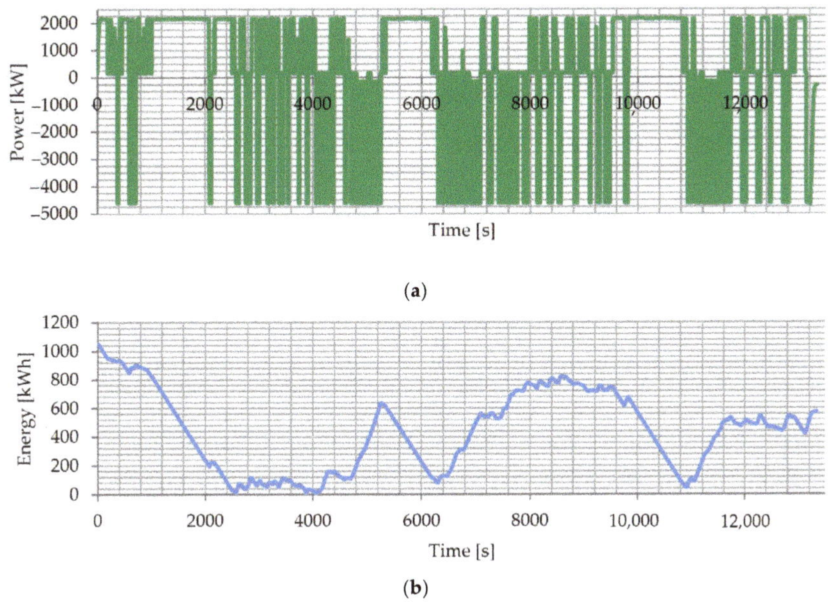

Figure 13. Time dependences of power (**a**) and energy change (**b**) of the accumulator when moving from the Kharkiv-Sortuvalnyi station to the Sumy station with loaded wagons.

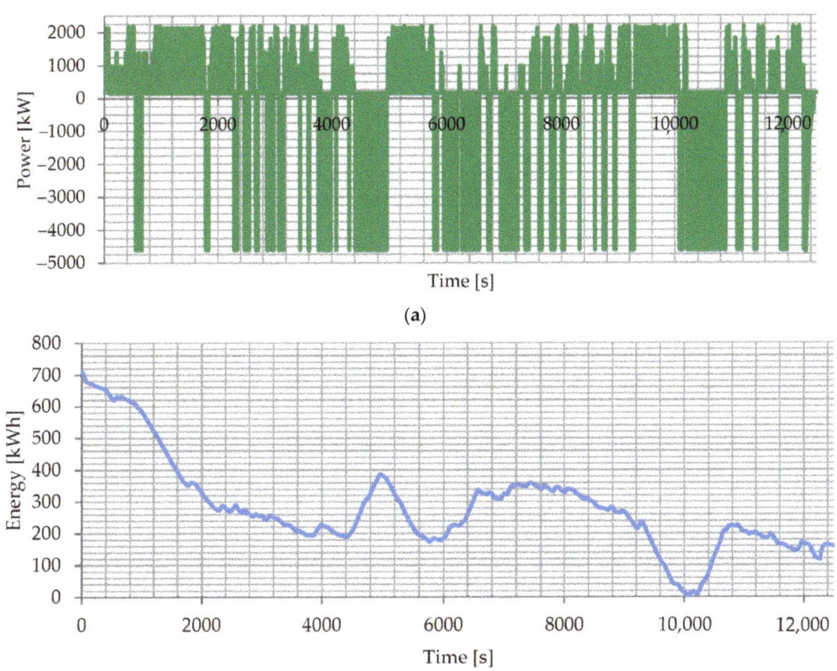

Figure 14. Time dependences of power (**a**) and energy change (**b**) of the accumulator when moving from Kharkiv-Sortuvalnyi station to Sumy station with empty wagons.

Thus, the use of an energy storage device to accumulate energy from the booster and diesel sections reduced fuel consumption during driving by half. At the same time, all energy that is available during electrodynamic braking is stored. The performance indicators of the diesel engine do not change.

4. Discussion of the Obtained Results

As a result of the conducted research, the effectiveness of the use of energy storage devices on the main freight diesel locomotive has been established. In order to determine a more effective organization of the traction system of a mainline diesel locomotive, a comparison of the obtained calculation results was made for the second (diesel section and booster section with energy storage) and the third (diesel section and booster section with energy storage with combined traction systems) options for organizing the traction drive of a mainline diesel locomotive.

As can be seen from Table 6, both cases of using the energy accumulator—for energy accumulation only from the booster section and accumulation from both the booster section and the diesel section—ensure a 50% reduction in fuel consumption compared to a serial diesel locomotive. The indicators that characterize the use of a diesel engine do not change. In both cases, when the train is moving with loaded wagons, the use of diesel engines based on the load can be considered effective. However, in terms of working time under load, diesel is only used for about half of the time. It is worth noting that these indicators are affected by the load algorithm of the diesel and booster sections, which, in this work, was assumed to be identical to the serial diesel locomotive. Therefore, in order to improve the use of diesel indicators, it is worth developing control algorithms that will ensure the improvement of diesel use, prioritizing the reduction in the duration of unproductive work.

Table 6. The results of the comparison of the parameters of the train movement, obtained according to options 2 and 3.

Compared Parameter	Accumulation only from the Booster Section				Accumulation from Both (Diesel and Booster) Sections			
Reduction in fuel consumption relative to the basic diesel locomotive, %	≈50				≈50			
Change in the utilization factor of the installed capacity, %	without change				without change			
Change in the diesel utilization rate over time, %	without change				without change			
Minimum working energy intensity, r. u.	1	1	1	1	0.495	0.637	0.445	0.597
Energy that must be stored in the accumulator before movement, r. u.	1	1	1	1	0.495	0.637	0.445	0.597
Power of the energy storage, r. u.	1				2			

The analysis of the indicators of the energy storage device shows that when storing energy only from the booster section, its power is approximately the same for both traction mode and storage mode. In the second case, the power in the storage mode is almost twice as high as the power in the energy return mode. At the same time, the working energy capacity for the second option is 1.57–2.25 less than for the first option.

For the normal functioning of the energy storage during movement, it must be initially charged. In the first case, using the storage device requires 2.25–1.57 times more energy, which must be stored in the storage device before the trip. That is, the total cost of fuel and energy resources for the trip will be higher in the first option of using the energy storage: before starting the movement, more energy must be accumulated in the storage. Charging can be carried out from a diesel section or—in case of long-term parking—from a specialized charging station [58,59].

When applying the energy storage according to the first variant of conversion, only one section of the diesel locomotive is subject to conversion. In the case of using the drive according to the second option, the conversion of both sections is necessary. In order to charge the energy storage, it is necessary to build appropriate infrastructure facilities [60–62]. In addition, the technological processes of stations, depots, and maintenance points should be optimized in order to ensure the charging of accumulators on the booster section and their diagnostics [63–65]. To increase the efficiency of the introduction of energy storage devices on the locomotive, it is necessary to develop energy-optimal management strategies for such locomotives [66–68].

Thus, with the investigated methods of using energy storage devices on the 2TE116 diesel locomotive, a two-fold reduction in fuel consumption during its movement on the route is achieved. At the same time, there is a need to charge the energy storage before starting movement, for which additional energy resources will be used. For the final choice of the energy storage option on the 2TE116 diesel locomotive, further research is required, taking into account the economic aspects of the diesel locomotive modernization and the creation of an infrastructure for servicing the energy storage.

5. Conclusions

Ways of improving the two-section mainline freight locomotive, in which energy storage is installed in one section instead of the diesel generator, are considered. The cases where the traction system of the section with energy storage is isolated from the traction system of the section with the diesel generator, as well as the case when the traction systems of both sections are connected, have been investigated. Through mathematical modeling, it was determined that in electrodynamic braking modes, depending on the configuration of the traction system, 32.64% of the energy spent in traction mode can be accumulated. The use of this energy in traction mode provides a reduction in fuel consumption when driving on the route by about 50%. To determine the final option of modernization, it is necessary to take into account the economic aspects that arise in connection with the introduction of an energy storage device during the modernization of a diesel locomotive.

Author Contributions: Conceptualization, I.R. and L.N.; methodology, S.G. and I.R.; software, S.G. and I.R.; validation, L.N. and I.R.; formal analysis, S.G. and I.R.; resources, I.R. and L.N.; data curation, I.R. and S.G.; writing—original draft preparation, I.R. and S.G.; writing—review and editing, L.N. and S.G.; visualization, I.R. and S.G.; supervision, I.R.; project administration, L.N. All authors have read and agreed to the published version of the manuscript.

Funding: This research received no external funding.

Data Availability Statement: Dataset available on request from the authors.

Conflicts of Interest: The authors declare no conflicts of interest.

References

1. Available online: https://transport.ec.europa.eu/transport-themes/infrastructure-and-investment/trans-european-transport-network-ten-t_en (accessed on 1 February 2024).
2. Kharchuk, O.; Kryvenko, O. Railway International Transport Corridors Ukraine and Role USCTS «Liski» in Their Development. *Int. Sci. J. Internauka* **2017**, *15*, 63–68. (In Ukrainian)
3. Available online: https://www.uz.gov.ua/press_center/up_to_date_topic/607214/ (accessed on 1 February 2024).
4. Smyrnov, I.; Lyubitseva, O.; Zapototskyi, S.; Hryniuk, D.; Stepanets, I. European Transport Corridors: Experience and Opportunities for Ukraine in Times of Danger. In *Smart Technologies in Urban Engineering. STUE 2023*; Lecture Notes in Networks and Systems, Arsenyeva, O., Romanova, T., Sukhonos, M., Biletskyi, I., Tsegelnyk, Y., Eds.; Springer: Cham, Switzerland, 2023; Volume 807. [CrossRef]
5. Available online: https://www.wabteccorp.com/locomotive/heavy-haul-locomotives/es44aci-locomotive (accessed on 1 February 2024).
6. Available online: https://logist.today/dnevnik_logista/2018-12-17/v-uz-rasskazali-kak-projavil-sebja-lokomotiv-trizub-v-pervyh-rejsah/ (accessed on 1 February 2024).
7. Available online: https://finance.bigmir.net/news/2852901-made-in-ukraine-kakie-poezda-mogut-zamenit-staryj-park-ukrzaliznyci (accessed on 1 February 2024).
8. Available online: https://trz.com.ua/modernization-ua/ (accessed on 1 February 2024).

9. Kljaić, Z.; Pavković, D.; Cipek, M.; Trstenjak, M.; Mlinarić, T.J.; Nikšić, M. An Overview of Current Challenges and Emerging Technologies to Facilitate Increased Energy Efficiency, Safety, and Sustainability of Railway Transport. *Future Internet* **2023**, *15*, 347. [CrossRef]
10. Cipek, M.; Pavković, D.; Kljaić, Z.; Trstenjak, M.; Mlinarić, T.J. Assessment of Battery-Hybrid Diesel-Electric Locomotive Fuel Savings and Emission Reduction Potentials Based on a Realistic Mountainous Rail Route. *Energy* **2019**, *173*, 1154–1171. [CrossRef]
11. Mayrink, S., Jr.; Oliveira, J.G.; Dias, B.H.; Oliveira, L.W.; Ochoa, J.S.; Rosseti, G.S. Regenerative Braking for Energy Recovering in Diesel-Electric Freight Trains: A Technical and Economic Evaluation. *Energies* **2020**, *13*, 963. [CrossRef]
12. Saeed, M.; Briz, F.; Guerrero, J.M.; Larrazabal, I.; Ortega, D. Onboard Energy Storage Systems for Railway: Present and Trends. *Open J. Ind. Appl.* **2023**, *4*, 238–259. [CrossRef]
13. Li, W.; Wang, C.; Pei, H.; Xu, C.; Lin, G.; Deng, J.; Jiang, D.; Huang, Y. An Improved Energy Management Strategy of Diesel-Electric Hybrid Propulsion System Based on FNN-DP Strategy. *Electronics* **2023**, *12*, 486. [CrossRef]
14. Meinert, M.; Prenleloup, P.; Schmid, S.; Palacin, R. Energy Storage Technologies and Hybrid Architectures for Specific Diesel-Driven Rail Duty Cycles: Design and System Integration Aspects. *Appl. Energy* **2015**, *157*, 619–629. [CrossRef]
15. Available online: https://www.wabteccorp.com/locomotive/alternative-fuellocomotives/flxdrive (accessed on 1 February 2024).
16. Sun, Y.; Cole, C.; Spiryagin, M.; Godber, T.; Hames, S.; Rasul, M. Conceptual Designs of Hybrid Locomotives for Application as Heavy Haul Trains on Typical Track Lines. *Proc. Inst. Mech. Eng. Part F J. Rail Rapid Transit* **2013**, *227*, 439–452. [CrossRef]
17. da Silva Moraes, C.G.; Brockveld, S.L.; Heldwein, M.L.; Franca, A.S.; Vaccari, A.S.; Waltrich, G. Power Conversion Technologies for a Hybrid Energy Storage System in Diesel-Electric Locomotives. *IEEE Trans. Ind. Electron.* **2021**, *68*, 9081–9091. [CrossRef]
18. Fedele, E.; Iannuzzi, D.; Pizzo, A.D. Onboard Energy Storage in Rail Transport: Review of Real Applications and Techno-Economic Assessments. *IET Electr. Syst. Transp.* **2021**, *11*, 279–309. [CrossRef]
19. Omelyanenko, V.I.; Riabov, I.S.; Overianova, L.V.; Omelianenko, H.V. Traction Electric Drive Based on Fuel Cell Batteries and On-Board Inertial Energy Storage for Multi Unit Train. *Electr. Eng. Electromechan.* **2021**, *4*, 64–72. [CrossRef]
20. Kapetanović, M.; Núñez, A.; van Oort, N.; Goverde, R.M.P. Reducing Fuel Consumption and Related Emissions Through Optimal Sizing of Energy Storage Systems for Diesel-Electric Trains. *Appl. Energy* **2021**, *294*, 117018. [CrossRef]
21. Mayet, C.; Pouget, J.; Bouscayrol, A.; Lhomme, W. Influence of an Energy Storage System on the Energy Consumption of a Diesel-Electric Locomotive. *IEEE Trans. Veh. Technol.* **2014**, *63*, 1032–1040. [CrossRef]
22. Shiraki, N.; Satou, H.; Arai, S. A Hybrid System for Diesel Railcar Series Ki-Ha E200. In Proceedings of the 2010 International Power Electronics Conference—ECCE ASIA, Sapporo, Japan, 21–24 June 2010; pp. 2853–2858. [CrossRef]
23. Fayad, A.; Ibrahim, H.; Ilinca, A.; Sattarpanah Karganroudi, S.; Issa, M. Energy Recovering Using Regenerative Braking in Diesel–Electric Passenger Trains: Economical and Technical Analysis of Fuel Savings and GHG Emission Reductions. *Energies* **2022**, *15*, 37. [CrossRef]
24. Barbosa, F.C. Battery Only Electric Traction for Freight Trains—A Technical and Operational Assessment. *Proc. Inst. Mech. Eng. Part F J. Rail Rapid Transit* **2023**, *238*, 322–337. [CrossRef]
25. Cipek, M.; Pavković, D.; Krznar, M.; Kljaić, Z.; Trstenjak, M.; Mlinarić, T.J. Comparative Analysis of Conventional Diesel-Electric and Hypothetical Battery-Electric Heavy Haul Locomotive Operation in Terms of Fuel Savings and Emissions Reduction Potentials. *Energy* **2021**, *232*, 121097. [CrossRef]
26. Iden, M.E. Battery Storage of Propulsion-Energy for Locomotives. In Proceedings of the 2014 Joint Rail Conference, Colorado Springs, CO, USA, 2–4 April 2014. [CrossRef]
27. Masliev, V. *Modern Designs and Dynamics of Railway Rolling Stock*; NTU "KhPI": Kharkiv, Ukraine, 2014. (In Ukrainian)
28. Zayonchkovskiy, V.; Masliev, V.; Yakunin, D. *The General Structure, Thermal Calculations of Power Plants and Issues of Manufacturing Technology for the Crew Part of Diesel Locomotives*; NTU "KhPI": Kharkiv, Ukraine, 2008. (In Ukrainian)
29. Tartakovsky, E.; Agulov, A.; Falendish, A.; Basov, G. *Theory and Design of Locomotives. Part 1. General Characteristics and Structure of Locomotives*; NTU "KhPI": Kharkiv, Ukraine, 2006. (In Ukrainian)
30. Available online: https://suspilne.media/273737-ukraina-i-moldova-vidnovili-alternativnij-zaliznicnij-marsrut-miz-odesou-ta-dunajskimi-portami/ (accessed on 1 February 2024).
31. Krushedolsky, A.; Grishina, E. Adjustment of Diesel Locomotive Characteristics of a Forced Diesel Engine of a Main Diesel Locomotive. *Intern. Combust. Engines* **2009**, *1*, 103–106. (In Ukrainian)
32. Bondarenko, I.; Severino, A.; Olayode, I.O.; Campisi, T.; Neduzha, L. Dynamic Sustainable Processes Simulation to Study Transport Object Efficiency. *Infrastructures* **2022**, *7*, 124. [CrossRef]
33. Bosso, N.; Gugliotta, A.; Zampieri, N. Design and Simulation of a Railway Vehicle for the Transport of People with Reduced Mobility. *Shock. Vib.* **2018**, *2018*, 9207639. [CrossRef]
34. Kalivoda, J.; Neduzha, L. Running Dynamics of Rail Vehicles. *Energies* **2022**, *15*, 5843. [CrossRef]
35. Ye, Z.; Chen, Z.; Zhou, Z.; Zhai, W. Dynamic Modelling of Locomotive with Direct Drive Traction System Considering Electromechanical Coupling Effect. *Veh. Syst. Dyn.* **2023**, *63*, 1–24. [CrossRef]
36. Holub, H.; Dmytrychenko, M.; Kulbovskyi, I.; Sapronova, S. Modeling of Energy-Saving Technologies in Traction Rolling Stock Projects. In Proceedings of the 7th ASRES International Conference on Intelligent Technologies. ICIT 2022, Jakarta, Indonesia, 16–18 December 2022; Lecture Notes in Networks and Systems. Arya, K.V., Tripathi, V.K., Rodriguez, C., Yusuf, E., Eds.; Springer: Singapore, 2023; Volume 683. [CrossRef]

37. Dižo, J.; Blatnický, M.; Steišūnas, S. Multibody Simulation of Rail Vehicle Running Considering Track Flexibility. In *TRANSBALTICA XI: Transportation Science and Technology*; TRANSBALTICA 2019. Lecture Notes in Intelligent Transportation and Infrastructure; Gopalakrishnan, K., Prentkovskis, O., Jackiva, I., Junevičius, R., Eds.; Springer: Cham, Switzerland, 2020. [CrossRef]
38. Sablin, O.; Bosyi, D.; Kuznetsov, V.; Lewczuk, K.; Kebal, I.; Myamlin, S.S. Efficiency of Energy Storage Control in the Electric Transport Systems. *Arch. Transp.* **2022**, *62*, 105–122. [CrossRef]
39. Yatsko, S.; Sytnik, B.; Vashchenko, Y.; Sidorenko, A.; Liubarskyi, B.; Veretennikov, I.; Glebova, M. Comprehensive Approach to Modeling Dynamic Processes in the System of Underground Rail Electric Traction. *East.-Eur. J. Enterp. Technol.* **2019**, *1*, 48–57. [CrossRef]
40. Zou, Z.; Zheng, M.; Lu, Q. Modeling and Simulation of Traction Power Supply System for High-Speed Maglev Train. *World Electr. Veh. J.* **2022**, *13*, 82. [CrossRef]
41. Bondarenko, I.; Lukoševičius, V.; Keršys, R.; Neduzha, L. Investigation of Dynamic Processes of Rolling Stock–Track Interaction: Experimental Realization. *Sustainability* **2023**, *15*, 5356. [CrossRef]
42. Li, L.; Dutta, S.; Dixon, R.; Stewart, E. Railway Track Switch Simulation: A New Dynamic Model for Studying Actuator and Switch Blade Dynamics. *Proc. Inst. Mech. Eng. Part F J. Rail Rapid Transit* **2023**, *237*, 775–783. [CrossRef]
43. Holub, H.; Kulbovskyi, I.; Skok, P.; Melnychenko, O.; Kharuta, V.; Bambura, O.; Tretynychenko, Y. System Model of Information Flows in Networks of the Electric Supply System in Transport Infrastructure Projects. In Proceedings of the Transport Means 2020: Proceedings of the 24th International Conference, Kaunas, Lithuania, 30 September–2 October 2020; pp. 132–135.
44. Musayev, J.; Zhauyt, A.; Ismagulova, S.; Yussupova, S. Theory and Practice of Determining the Dynamic Performance of Traction Rolling Stock. *Appl. Sci.* **2023**, *13*, 12455. [CrossRef]
45. Gołębiowski, P.; Góra, I.; Bolzhelarskyi, Y. Risk Assessment in Railway Rolling Stock Planning. *Arch. Transp.* **2023**, *65*, 137–154. [CrossRef]
46. Kyrychenko, H.; Strelko, O.; Berdnychenko, Y.; Berdnychenko, I. Development of Automation of Operational Work of Railway Stations of Ukraine. In Proceedings of the 2023 IEEE International Conference on Information and Telecommunication Technologies and Radio Electronics (UkrMiCo), Kyiv, Ukraine, 13–18 November 2023; pp. 71–75. [CrossRef]
47. Slashchev, V. *Traction and Braking Calculations on Rail Transport*; Eastern Ukrainian National University Named after V. Dalya: Luhansk, Ukraine, 2005. (In Ukrainian)
48. Bobyr, D.; Grischenko, N.; Serdyuk, V.N. *The Theory of Locomotive Traction*; Educational and Scientific Institute "Dnipro Institute of Infrastructure and Transport": Dnipro, Ukraine, 2022. (In Ukrainian)
49. Riabov, I.; Goolak, S.; Kondratieva, L.; Overianova, L. Increasing the Energy Efficiency of the Multi-Motor Traction Electric Drive of an Electric Locomotive for Railway Quarry Transport. *Eng. Sci. Technol. Int. J.* **2023**, *42*, 101416. [CrossRef]
50. Syrotenko, Y. Determining the Power of a Shunting Diesel Locomotive Taking into Account the Place of its Operation. *East. -Eur. J. Enterp. Technol.* **2014**, *1*, 41–45. Available online: http://nbuv.gov.ua/UJRN/Vejpte_2014_1(8)_9 (accessed on 1 February 2024). (In Ukrainian).
51. Buriakovskiy, S.; Maslii, A.; Pomazan, D. Searching of The Optimum Configuration of The Traction Electric Transmission of The Shunting Locomotive. In Proceedings of the 2019 IEEE International Conference on Modern Electrical and Energy Systems (MEES), Kremenchuk, Ukraine, 23–25 September 2019; pp. 22–25. [CrossRef]
52. Goolak, S.; Riabov, I.; Tkachenko, V.; Sapronova, S.; Rubanik, I. Model of Pulsating Current Traction Motor Taking into Consideration Magnetic Losses in Steel. *Electr. Eng. Electromechan.* **2021**, *6*, 11–17. [CrossRef]
53. Afanasov, A.M.; Shapovalov, O.S.; Holik, S.N.; Arpul, S.V.; Bilukhin, D.S. Energy efficiency of heat tests for traction electric machines. *IOP Conf. Ser. Mater. Sci. Eng.* **2020**, *985*, 012026. [CrossRef]
54. Bodnar, B.; Kapitsa, M.; Bobyr, D.; Kyslyi, D. Defining the limits of application and the values of integration variables for the equations of train movement. *Nauk. Visnyk Natsionalnoho Hirnychoho Universytetu* **2019**, *6*, 59–65. [CrossRef]
55. Pugi, L. Synergic Design and Simulation of Battery-Operated Trains on Partially Electrified Lines: A Case Study regarding the Firenze Faenza Line. *Energies* **2024**, *17*, 24. [CrossRef]
56. Buriakovskiy, S.; Maslii, A.; Pomazan, D.; Panchenko, V.; Overianova, L.; Omelianenko, H. Multi-criteria Quality Evaluation of Energy Storage Devices for Rolling Stock Using Harrington's Desirability Function. In Proceedings of the IEEE 7th International Conference on Energy Smart Systems (ESS), Kyiv, Ukraine, 12–14 May 2020; pp. 158–163. [CrossRef]
57. Kondratieva, L.; Bogdanovs, A.; Overianova, L.; Riabov, I.; Goolak, S. Determination of the working energy capacity of the on-board energy storage system of an electric locomotive for quarry railway transport during working with a limitation of consumed power. *Arch. Transp.* **2023**, *65*, 119–135. [CrossRef]
58. Noga, M. *Research on the Possibility of Using a Hybrid Electric Vehicle Powertrain in V2X Applications*; SAE Technical Paper 2022-01-1134; SAE International: Warrendale, PA, USA, 2022. [CrossRef]
59. Guov, V.; Jackson, D.; Obi, M.; Cao, Y. Sizing BESS and On-site Renewable for Battery-electric Freight Rail Charging Station. In Proceedings of the 2022 IEEE 13th International Symposium on Power Electronics for Distributed Generation Systems (PEDG), Kiel, Germany, 26–29 June 2022; pp. 1–6. [CrossRef]
60. Leijon, J.; Boström, C. Charging Electric Vehicles Today and in the Future. *World Electr. Veh. J.* **2022**, *13*, 139. [CrossRef]
61. Aredah, A.; Du, J.; Hegazi, M.; List, G.; Rakha, H.A. Comparative analysis of alternative powertrain technologies in freight trains: A numerical examination towards sustainable rail transport. *Appl. Energy* **2024**, *356*, 122411. [CrossRef]

62. Ruvio, A.; Bayrak, O. A preliminary design of a hybrid train's on-board batteries for a 25 kV–50 Hz high speed railway line. *J. Energy Storage* **2024**, *84 Pt B*, 110966. [CrossRef]
63. Kulbovskyi, I.; Holub, H.; Kyiashko, V.; Skliarenko, I.; Sorochynska, O.; Gurenkova, O.; Kharuta, V. Investigation of the model of functioning of production and technological potential of subway power supply departments. In Proceedings of the Transport Means 2020: Proceedings of the 24th International Conference, Kaunas, Lithuania, 30 September–2 October 2020; pp. 416–420.
64. Gubarevych, O.; Goolak, S.; Melkonova, I.; Yurchenko, M. Structural Diagram of the Built-in Diagnostic System for Electric Drives of Vehicles. *Diagnostyka* **2022**, *23*, 2022406. [CrossRef]
65. Urbaniak, M.; Kardas-Cinal, E.; Jacyna, M. Optimization of Energetic Train Cooperation. *Symmetry* **2019**, *11*, 1175. [CrossRef]
66. Cipek, M.; Pavković, D.; Kljaić, Z. Optimized Energy Management Control of a Hybrid Electric Locomotive. *Machines* **2023**, *11*, 589. [CrossRef]
67. Buriakovskyi, S.; Asmolova, L.; Obruch, I.; Maslii, A.; Pomazan, D. Simulation Modeling of a Hybrid Diesel Locomotive Power Plant Based on a Free-Piston Engine. In Proceedings of the 2022 IEEE 4th International Conference on Modern Electrical and Energy System (MEES), Kremenchuk, Ukraine, 20–23 October 2022; pp. 1–4. [CrossRef]
68. Afanasov, A.; Linik, D.; Arpul, S.; Belukhin, D.; Vasylyev, V. Prospects of Using Autonomous Electric Trains with Onboard Storage Stores. *Transp. Syst. Transp. Technol.* **2022**, *23*, 46–51. [CrossRef]

vehicles

MDPI

Article

A New Strategy for Railway Bogie Frame Designing Combining Structural–Topological Optimization and Sensitivity Analysis

Alessio Cascino *, Enrico Meli and Andrea Rindi

Department of Industrial Engineering, University of Florence, 50139 Florence, Italy; enrico.meli@unifi.it (E.M.); andrea.rindi@unifi.it (A.R.)
* Correspondence: alessio.cascino@unifi.it

Abstract: Rolling stock manufacturers are finding innovative structural solutions to improve the quality and reliability of railway vehicles components. Structural optimization processes represent an effective strategy for reducing manufacturing costs, resulting in geometries easier to design and produce. In this framework, the present paper proposes a new methodology to design a railway metro bogie frame, combining structural–topological optimization methods and sensitivity analysis. In addition, manufacturing constraints were included to make the component design suitable for production through sand-casting. A robust sensitivity analysis has highlighted the most critical load conditions acting on the bogie frame. Its effectiveness was verified by carrying out two different structural optimizations based on different loadings. Two equivalent designs were obtained. Computational times were positively reduced by about 57%. The maximum value of stress was reduced about 23%. This new methodology has shown encouraging results to streamline the design process of this complex mechanical system, allowing researchers to also include manufacturing requirements.

Keywords: railway vehicles; structural optimization; topology optimization; sensitivity analysis; manufacturing constraints; railway components; railway bogie

1. Introduction

Today, railway vehicles stand out as one of the optimal solutions for sustainable mobility, aiming to reduce air pollution generated by various types of land and non-land vehicles. The growth of the railway sector needs to address new challenges in both vehicle design and management. From a mechanical design perspective, one of the primary focuses is to standardize and make complex components such as a bogie frame increasingly replicable while allowing for easy customization based on the designer needs. This is why the use of diverse manufacturing techniques, such as sand-casting, supported by advanced numerical simulations, can lead to the realization of innovative and effective designs in rapid and acceptable timeframes. The ability to have a replicable bogie frame model would result in a significant improvement in production times, both for initial construction and in the event of damage and the subsequent need for component replacement. To achieve these goals, topological–structural optimization processes represent a fundamental tool supporting the design and development phase of the system under consideration. The available literature illustrates that not many studies about optimization processes, applied on railway vehicles' bogie design, have been carried out. This aspect is even more accentuated for metro vehicles, the object of the present work. Generally, topological optimization and structural optimization processes represent effective techniques to design several classes of components. A control arm used in automotive applications has been optimized to reduce its mass [1,2]. A high-performance chassis design of an automotive vehicle body has been innovated using topological optimization [3]. Similar approaches have been used to improve the design procedure for turbomachinery components, especially if combined with additive manufacturing techniques [4–8]. Referring to railway applications, the anchor bracket has been redesigned by taking advantage of a topology optimization process and introducing composite materials [9].

Many researchers have optimized the car body oriented towards a lightweight design process [10–12].

Size optimization and modal behavior of the car body structure have been combined to propose a new design procedure for the car body [13,14].

A major part of the applications that involve manufacturing constraints are referred to as general applications. A Heaviside parametrization design to obtain manufacturable cast geometries adopting a gradient driven topology optimization approach is reported in [15]. In [16], a comparison between two casted parts designed through optimization process, with and without manufacturing constraints, is presented. A structural optimization approach focused on the bogie frame of a freight wagon and its bolster is described in [17]. The objective was a mass reduction in the system. A topological optimization was conducted using the SIMP method to maximize the stiffness of the structure. The bogie frame of a tilting train for Korean railways has been redesigned exploiting an optimization approach based on a microgenetic algorithm [18].

Furthermore, the literature highlights the importance of railway bogie frame life prediction. Multibody dynamics and a finite element approach have been combined to simulate the behavior of bogie frames under real operational service loads [19]. A rigid–flexible coupled vehicle dynamic model has been adopted to estimate the dynamic response acting on a high-speed train [20]. Similar approaches, applied on different types of bogie frame, have been proposed by other researchers [21–23]. Lifetime assessment and optimization have been applied on a welded a-type frame for a mining truck [24]. Experimental fatigue evaluation has been carried out on a metro bogie frame [25,26]. Stress and acceleration tests on the bogie frame of a diesel multiple unit (DMU) have been performed to clarify the reasons for fatigue cracking [27]. However, all of the proposed techniques aimed to assess the performance of the vehicle without considering component redesign and innovation.

In this research activity, the frame of a railway bogie used for a metropolitan vehicle was studied to create an innovative design. A robust campaign of topological–structural optimization was conducted on the component with the aim of achieving an appropriate simulation setting. Two main aspects characterized the novelty of this research. The first involved the utilization and control of technological constraints oriented towards an innovative and potentially manufacturable design for sand-casting. This approach is yet to be explored in the railway field, especially for critical components such as railway bogie frames. The second innovation included the implementation of a comprehensive sensitivity analysis to understand the most critical load conditions used for the mechanical verification of this component. This allowed for a streamlined and focused structural optimization based on the sensitivity analysis results. Significant reductions in computational times were achieved. The obtained design proved to be consistent and practically identical to that obtained through the complete simulation, fully meeting the objectives of this research activity. In addition, improvements in terms of stress distributions and concentrations were observed.

This paper is organized in four sections, including the introduction of Section 1. The bogie frame used as benchmark, the proposed methodology and the main modeling steps are described in Section 2. Section 3 reports all the results and discussions about sensitivity analysis, structural optimization, static and fatigue structural verification and frame redesign. Finally, in Section 4, the conclusions and future developments of the activity are presented.

2. Materials and Methods

In this section, the methodology that was followed during the present research activity is presented. First, a brief presentation of the metro bogie frame is reported. Then, the structural optimization process and the sensitivity analysis approach are explained in detail.

2.1. Model Description: The Metro Bogie Frame

The metro bogie frame is a crucial structural component designed to support and integrate various elements within the train undercarriage. The robust frame provides the necessary strength, durability, and resilience required for the demanding operational conditions of a metro system. The frame is meticulously made through precision welding techniques, ensuring a seamless and uniform connection of steel components. However, welded joints represent the main critical aspect of these complex mechanical systems, especially from the mechanical fatigue point of view. The frame consists of interconnected beams, cross-members, and other geometric configurations strategically designed to distribute the load evenly across the structure. This design not only enhances the bogie stability but also contributes to the overall safety and efficiency of the metro system. The structure of the original bogie frame was made up with construction steel and assembled through welding, as described above. Figure 1 illustrates a complete view of it.

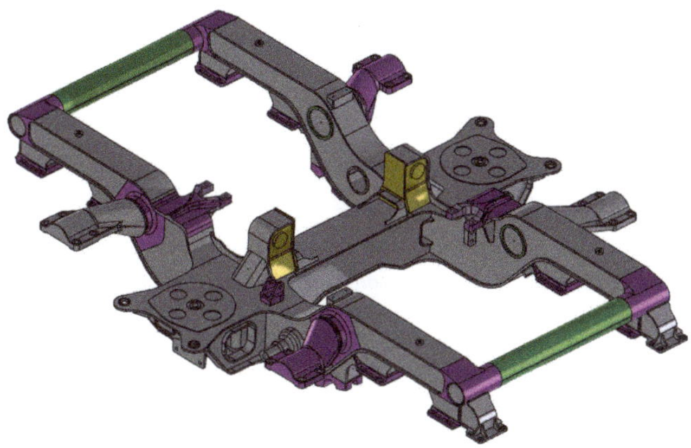

Figure 1. View of the original metro bogie frame.

The perimeter crossbeams (green color in Figure 1) had a circular section and provided a significant contribution to the frame resistance against all torsional loads. Their end elements were directly welded to the main longitudinal members. The device for supporting the brake components, as shown in Figure 2, consisted of an unalloyed steel element that was laterally welded to the outer surfaces of the longitudinal members in an area near the primary suspension.

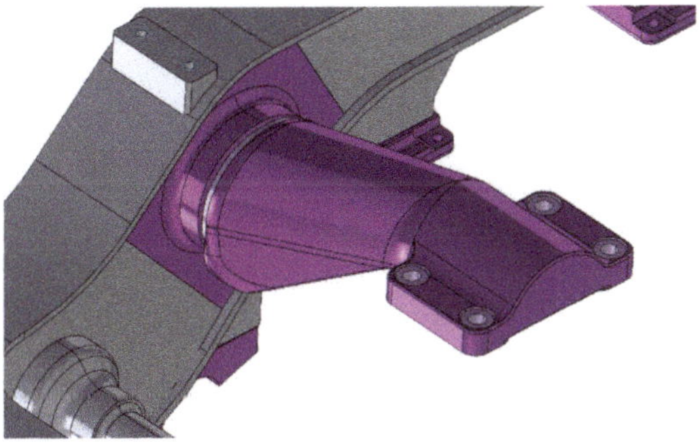

Figure 2. View of the original braking support system.

The support for the traction link was a rotational component initially fitted onto the longitudinal member through a hole and subsequently welded to the structure peripherally. It was important to highlight the hollow structure underneath the secondary suspension housing, designed to facilitate the passage of the traction link, properly connected to the bolster beam at the other end. The components described are clearly visible in Figure 3.

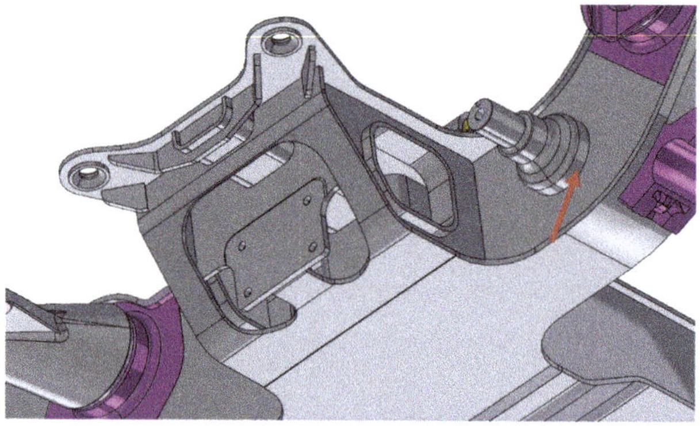

Figure 3. View of the original traction link zone.

Other crucial elements included the support for the gearbox system, located on the shoulder of the longitudinal beams and properly welded to its upper flange. The engine support was positioned near the traction link. Finally, the last component essential for understanding the force exchange is a welded element near the housing of the secondary suspension, responsible for transmitting transverse forces from the cylinder connected to the bolster beam (visible in Figure 1 and colored purple).

2.2. Methodology

The methodology adopted by the authors for the innovation of the present metro bogie frame is briefly described and summed up in Figure 4. It aims to combine the CAD environment, FE calculation and a structural optimization process which involves technological constraints oriented to the sand-casting process. One important objective of the activity was to create a suitable bogie frame design to produce it with a sand-casting process. It represents a real innovation in the railway field, where this type of mechanical system, due to its complex features, it is always made up of structural steel and assembled through a welding process. In order to assess the performance of the final design, the real performance of a material potentially suitable was included. In detail, the reference material was EN-GJS-450-10, the mechanical characteristics of which can be found in [28]. The main innovative approach of this research work involved the implementation of a robust sensitivity analysis of the loading conditions to which railway bogies are subjected. The primary objective was to expedite the complex structural optimization processes, which can take up to consecutive days of computational time. Secondary, the optimization process must converge to the same topology result to confirm the effectiveness of the sensitivity analysis. Then, two classes of optimizations were carried out to find the final design of the bogie, comparing mass distribution and calculation time.

First, the metro bogie frame was tested numerically according to the load conditions listed in reference standard EN 13749:2021 [29]. In detail, 20 load cases were tested, including static and fatigue load conditions. These are listed and described in Table 1.

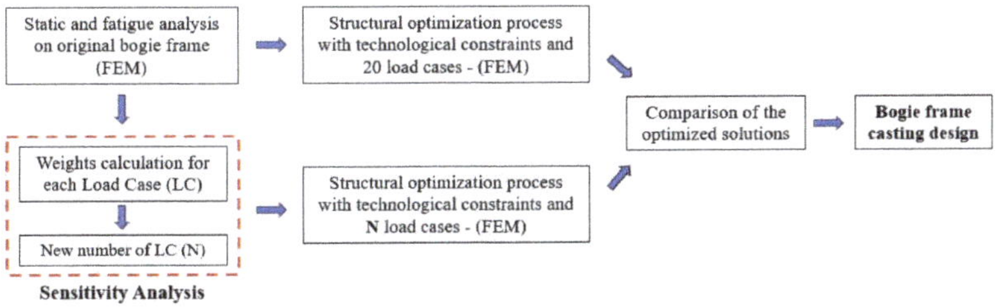

Figure 4. Main steps of the methodology.

Table 1. Load case conditions.

LC	Description
S01A	Lifting on four points of the frame
S02	Longitudinal force at the bogie pivot + vertical force on the secondary suspensions
S03	Lifting on two points of the frame (arranged diagonally)
S04	Main loads, operative loads
PR01	Main loads, operative loads, main systems mass
V04	Combination of loads applied only on the main supports
V05	Exceptional load under truck twist
PR10	Max load on secondary suspension with constraint on three wheels
F1	Main loads, operative loads, internal pressure of the air springs and truck twist combined
F2	Main loads, operative loads, internal pressure of the air springs and truck twist combined
PR02	Main loads and operative loads combined
PR03	Main loads and operative loads combined
PR04	Vertical inertial loads on the masses attached to the truck frame
PR05	Vertical inertial loads on the masses attached to the truck frame
PR06	Transversal inertial loads on the masses attached to the truck frame
PR07	Transversal inertial loads on the masses attached to the truck frame
PR08	Longitudinal inertial loads on the masses attached to the truck frame
PR09	Longitudinal inertial loads on the masses attached to the truck frame
PR11	Longitudinal lozenging forces
PR12	Longitudinal lozenging forces

Each load case was composed of several contributors. For this reason, all of the descriptions cite the main forces involved in the single case, without entering formulations and values, which is not useful for the comprehension of the present work. All the load cases between S01A and PR10 are static load cases, while others are used to test fatigue conditions.

2.3. Finite Element Model of the Bogie Frame

The finite element model of the starting bogie frame was a high-level representation of the system, made with an FE commercial software. Figure 5 illustrates it and a meshing detail in the most critical area. The structure of the component, including the central crossbeam, the longitudinal beam, and the perimeter crossbeams, was modeled using a two-dimensional mesh consisting of SHELL elements of QUAD4 type, with an average size of 10 mm and a minimum size of 3 mm. The sensitivity analysis performed on the component grid allowed us to ensure the good quality of the mesh and reliable results, especially for stress evaluation. All the supports described in the previous section, on the other hand, were composed of a three-dimensional mesh of second-order TETRA elements (TETRA10 type), with the same dimensions as the two-dimensional elements. The choice of different types of elements depended on the characteristics of the components. The bogie frame featured thinner thicknesses well represented by the numerical formulation that describes the mechanical behavior of shell elements. On the other hand, the supports had three-dimensional development geometries that necessitated solid elements. These two types of mesh could not be connected directly, as the 2D mesh had elements of the first order. An effective solution was found to join to the main structure all the supports modeled with solid mesh: the use of a freeze/bonded contact. This type of contact configuration ensures the linearity of the model, and the complete redistribution of all the stresses acting (both forces and moments). In detail, it enforces zero relative motion on the contact interfaces, the gap between them remains fixed at the original value and the sliding distance is forced to be zero. In this way, the use of the FE model was improved, ensuring the accuracy of the results and simplicity of modeling. Another important advantage of this approach was to be able to separately modify both the supports of the various subsystems and the main frame, increasing the speed of potential changes and subsequent numerical tests. Loads were applied with different approaches, according to the application zone. In addition to classical forces and pressure, zero-dimensional RBE3 elements were also utilized. These elements prevented the addition of stiffness to the structure, effectively distributing the loads among the respective nodes. The primary suspension (yellow color), which acted as a suspensive stage between the wheelset and the bogie frame, was modeled using one-dimensional elements. The accurate representation of this system was of crucial importance as it was essential to faithfully replicate this specific connection point, avoiding the introduction of excessively stiff constraints that could generate extremely high stresses, which did not result in realistic outcomes. The modeling of the axle (light blue color) and the pin seat was accomplished using one-dimensional beam elements with congruent sections. The conical rubber–metal suspension was simulated by employing two one-dimensional spring elements connected through two rigid multi-node elements of type RBE3 to the supports where the suspensions were located. Constraint conditions, for all load cases, were represented by an isostatic configuration, as reference points of the four wheels, positioned at the end of the axle. The model consisted of 572′513 elements and 692′502 nodes.

2.4. Sensitivity Analysis

Sensitivity analysis is a fundamental analytical process employed across various scientific domains with the aim of understanding how variations in the values of parameters within a complex system or mathematical model impact the outcomes or performances of that system. The process examines how variations in input data can propagate through these parameters, causing changes in the results achieved. Generally, this methodology consists of five fundamental steps: quantifying the uncertainty in each input, varying the parameters, identifying the model outputs to be analyzed, simulating the model a number of times, and calculating the sensitivity coefficients. The sensitivity study was conducted on the 20 load cases applied to the original bogie frame. The aim of this analysis was to understand how variations in load cases influenced the structural performance of the bogie, thereby identifying the most significant load scenarios. As the first step, the selected output were the stresses calculated at 6 critical points of the structure, identified following static and fatigue analyses. The choice of these points aimed to comprehensively represent the mechanical behavior of the entire bogie. Depending on the considered load case, the structure experienced varying stresses, and stress concentrations were observed in potentially distinct areas. In this regard, the chosen nodes allowed us to account for these effects. Figure 6 approximately shows the position of the six critical nodes.

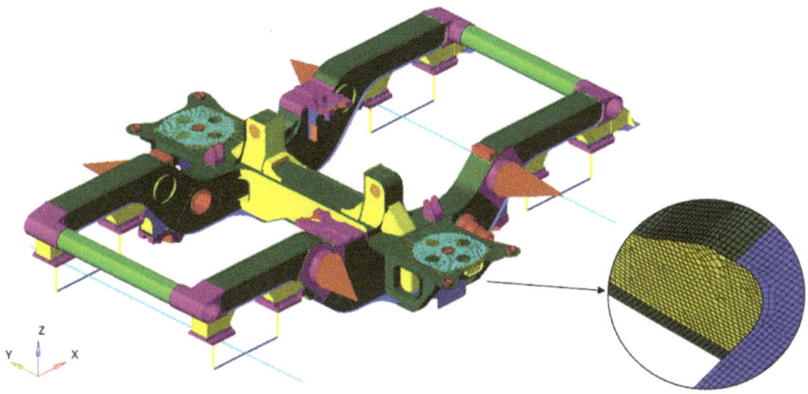

Figure 5. Finite element model of the starting bogie frame.

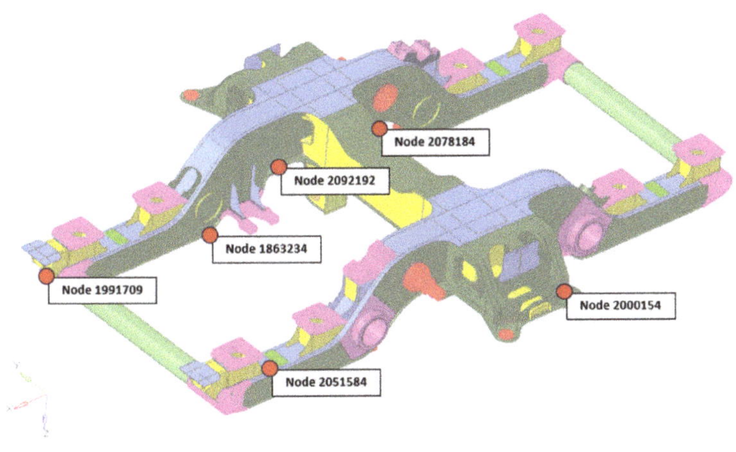

Figure 6. Critical nodes selected for the sensitivity analysis.

The second step involved quantifying the uncertainty for each chosen output. All load cases were analyzed, remembering that each was composed of a combination of a certain number of loads. Each contribution was tested at 25%, 50%, 75%, and 100% of the value required by the reference standard. This approach allowed for the consideration of realistic variations in the loading conditions to which the bogie frame could be statistically subjected, as well as verifying the linear behavior of the calculated outputs. It was demonstrated that by varying the percentage of each load contribution, stresses would vary linearly, thereby maintaining the criticality of the considered load cases unchanged. This analysis was conducted on all critical nodes, and the results are reported in Section 3.1. The sensitivity coefficients allowed us to quantify the importance of input parameters and understand how variations in these parameters propagated into the model results. To assess the uncertainty of the inputs, the normalized index was chosen. This was a dimensionless measure that standardized the input parameters so that they were comparable. In practice, it expressed the variation in each parameter relative to its range or maximum value. Expression (1) provides its definition.

$$I_{ij} = \frac{\sigma_{ij}}{max_j(\sigma_{ij})} \tag{1}$$

where the index i indicates the various critical points used in the sensitivity study and index j indicates the different load scenarios. I_{ij} is the local normalized index for node i and load scenario j, σ_{ij} is the stress in the critical node i calculated for the load scenario j, and $max_j(\sigma_{ij})$ represents the maximum stress for the single critical node i, varying j. Subsequently, a new index representing each load case at a global level was calculated. It was referred to as the Global Index (G_j). The new index was computed as the sum of the normalized values recorded at critical nodes for each load case. This approach allowed us to highlight the overall effect of parameter variations on all outputs, thereby identifying the most influential cases affecting the original bogie frame. Expression (2) describes it. All of the results related to this index are reported in Section 3.1.

$$G_j = \sum_i I_{ij} \tag{2}$$

2.5. Topological Optimization Settings

The topological optimization process conducted on the metro bogie frame during the present study served a dual purpose: to demonstrate the effectiveness of the sensitivity analysis performed under the loading conditions prescribed by the reference standard for component verification and to propose an innovative design for the frame. This design aimed to feature shape and geometry characteristics oriented to its realization through a sand-casting process. Technological constraints were introduced to guide the solver towards this solution in alignment with the second objective. The optimization process was executed using commercial software employing the gradient method as the solving algorithm. The user defined all input parameters for the optimization analysis, while the solution process was autonomously managed by the software. The solution was based on the Gradient-based Optimization Method [30].

The main difference between the two topological optimizations was in the number of load cases upon which these were based on. Once completed, a comparison in terms of material distribution and calculation times was carried out. It was fundamental to establish the effectiveness of sensitivity analysis on the load cases selection. Regarding the settings of the optimization processes, they are summed up in Table 2.

Table 2. Optimization settings.

	Optimization Settings	
General settings	Optimization objective	Weighted compliance minimization
	Optimization constraint	Mass fraction < 0.4
	Pattern grouping	Symmetry with respect to a plane (y normal)
Technological settings	Minimum feature size	25 mm
	Draw single	Removing material along Z direction

The optimization process had the objective of minimizing the weighted compliance calculated on all the load cases. The compliance is the sum of the strain energy of the mechanical model. The higher the value, the more the model is deformed under the same load. This essentially defines an inverse measure of stiffness, allowing the formulation of a more efficient minimum problem. As illustrated in Figure 7, before starting the procedure, the model of the bogie was adapted to be processed. With the aim of providing greater freedom to the optimizer for defining a new bogie design, the material volume of the frame was increased and geometrically simplified, using simple geometries and volumes. The dimensions were kept within an acceptable range defined by the mechanical component dimensions. The partitioning of design and non-design space was fundamental, allowing for the definition of the portion of material on which the solver could act. In Figure 7, the design space is shown in light blue, while the non-design space is represented in red. The non-design space mainly included areas of interface with other systems, load application zones, and connection zones to supports. Giving special attention to the latter, this choice allowed for the main interfaces to remain unchanged, enabling the separate management of modifications to the main frame and those to the supports.

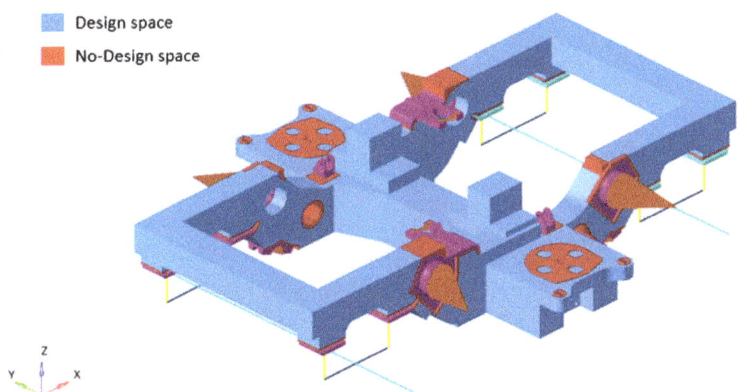

Figure 7. Topological optimization base model: design and non-design space.

The optimization constraint was imposed on mass fraction. It is the ratio between the final and the initial mass of the design space. It was set below 40%, a value generally adopted for topological optimization. The "Pattern grouping" setting allowed for the introduction of a symmetry condition with respect to a midplane of the bogie frame with the normal orientation along the y-direction. This condition was needed due to the characteristics of the loads that were applied only on one side of the bogie, creating an asymmetrical distribution of the material. After several tests, the value of the "Minimum feature size" setting was defined equal to 25 mm. It allowed us to ensure the correct level of detail in the optimized geometry. Finally, the "Draw single" setting forced the removal of material along a predefined direction. This technological constraint aims to simulate the extraction of the model from a mold, creating a geometry opened in the upper side or lower side, depending on the direction set.

3. Results and Discussion

3.1. Sensitivity Analysis: Uncertainty Quantification and Global Index

The uncertainty on the output used for the sensitivity analysis was assessed by varying all the load cases in three main percentages with respect to the full value: 75%, 50%, and 25%. This approach allowed us to check the linearity of the response and to evaluate realistic changes in the load. The responses were the stresses calculated in the six reference nodes. Figure 8 illustrates the results of the calculation carried out on node 2078184. The linearity of the outputs was confirmed.

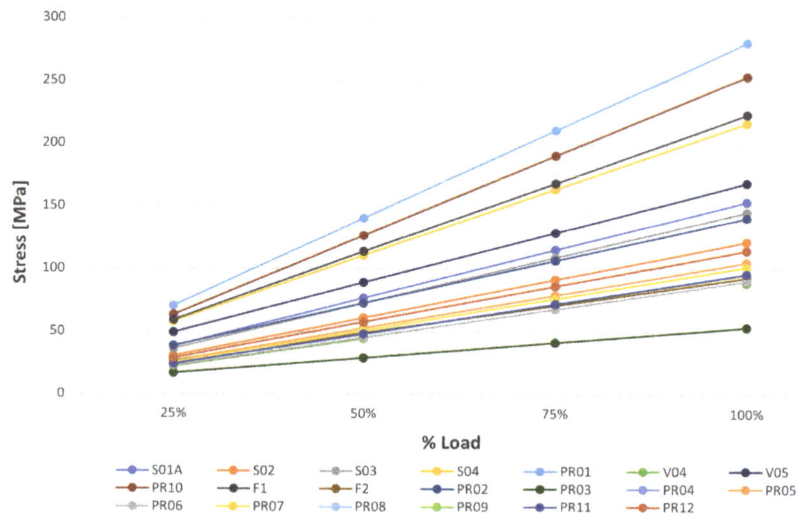

Figure 8. Output linearity assessment on node 2078184.

Consequently, the Global Index was assessed for all loading conditions on all nodes. The analysis showed seven critical load cases, with a Global Index value significantly higher with respect to the other ones, as reported in Figure 9.

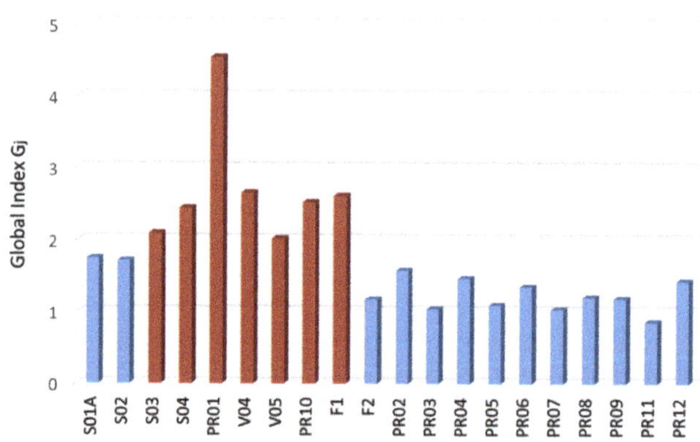

Figure 9. Global Index calculation result.

This result was fundamental to understand the most aggressive load conditions acting on a railway bogie frame. PR01 resulted in the most critical load, highlighting the important contribution of the suspended masses in terms of stress. V04 included the combination of loads acting on the main support interfaces, confirming the importance of these types of load. Finally, the F1 condition involved the effect of the internal pressure of the secondary suspension and of the truck twist. These, combined with the operative loads, resulted critical for the structure. Once the sensitivity analysis was completed, the second step of topological optimization was carried out.

3.2. *Topological Optimizations: Results and Comparison*

As shown in the dedicated methodology section, two topological optimizations were performed with the aim of evaluating the effectiveness of the sensitivity analysis conducted on the railway bogie frame. All optimization analysis settings remained unchanged. The only difference was in the number of load cases that the solver had to analyze: 20 in the full case and only 7 in the reduced case due to the sensitivity analysis. Both simulations were conducted using the same computer, which had the following characteristics: Intel(R) Xeon(R) CPU E5-2643 v4 @ 3.40GHz, RAM 32GB.

The results were analyzed based on two variables: element density distribution and computational times. Starting with the former, Figure 10a,b illustrates how the two distributions were highly consistent. Only minor differences were observed on the two lateral surfaces of the central crossbeam, with no consequential impact on the reconstruction of the final geometry. Material in that area would still be necessary due to the presence of attachments from various subsystems. Therefore, the two results could be considered effectively equivalent. Moving to computational times, a reduction of approximately 57% was observed between the full case and the reduced case, confirming the effectiveness of the proposed methodology.

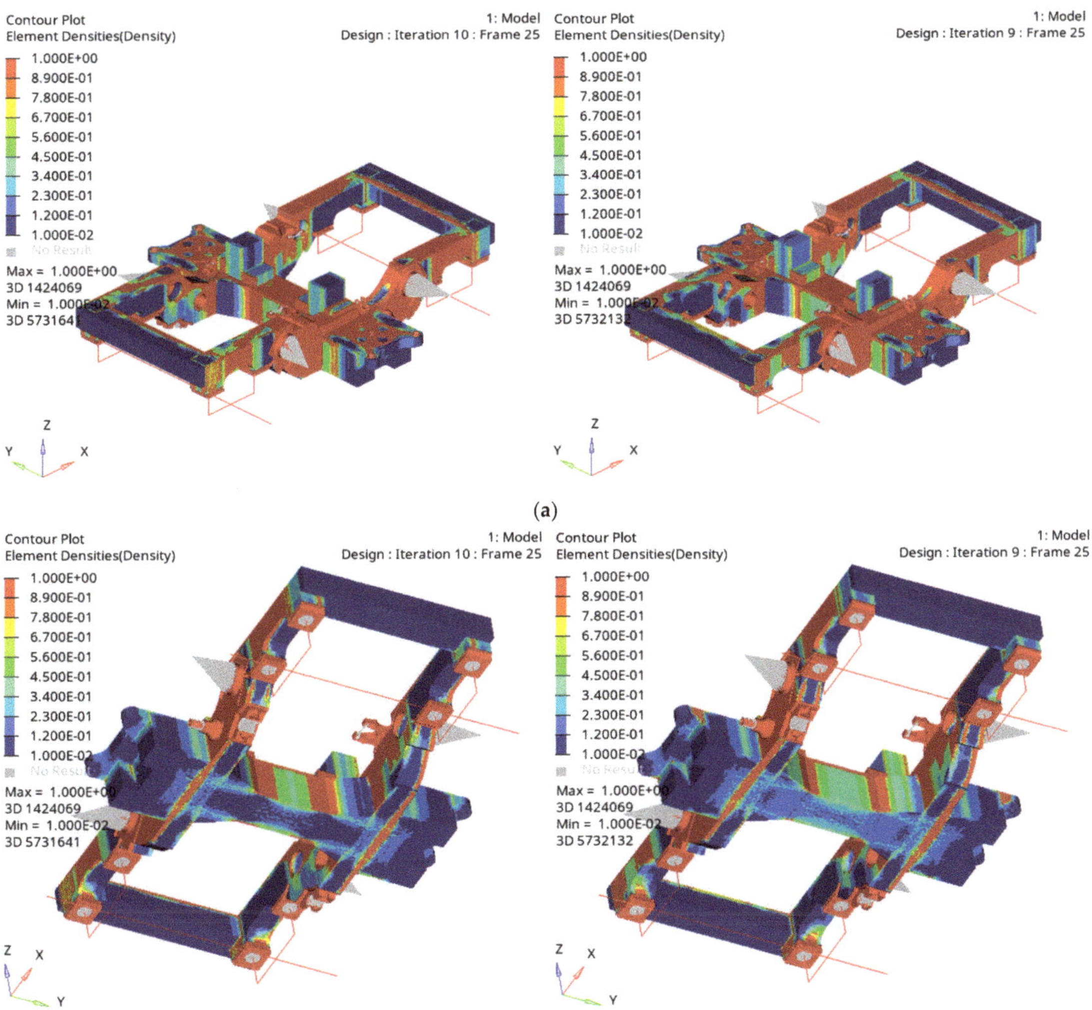

Figure 10. (**a**,**b**) Element density distribution: complete analysis vs. reduced analysis.

3.3. Bogie Frame: Innovative Design and Verification

Once the optimization process was completed, the obtained result was used as a basis upon which to build the final design of the bogie frame. Manufacturing constraints allowed us to have a geometry already compliant with the sand-casting manufacturing process. The reconstruction began with the central cross beam, and then moved on to the two longitudinal beams, concluding with the details related to the supports and various interface zones. The structure was open in the lower area, allowing for the extraction of the main frame structure without the use of cores, which would have been very costly due to the size of the frame. Certainly, in this manufacturing process, some cores would be required around the secondary suspensions, as shown in the subsequent figures. In order to make the structure stiffer and more resistant, both in terms of flexural and torsional aspects, reinforcement sets were introduced throughout the lower area of the frame. Additionally, these were crucial to ensure as uniform a flow of molten material as possible. Furthermore, having an open structure downward, towards the rail, prevents the accumulation of external agents that could damage the metal structure. Figure 11a illustrates the final design obtained for the crossbeam, while Figure 11b shows that for the secondary suspension zone.

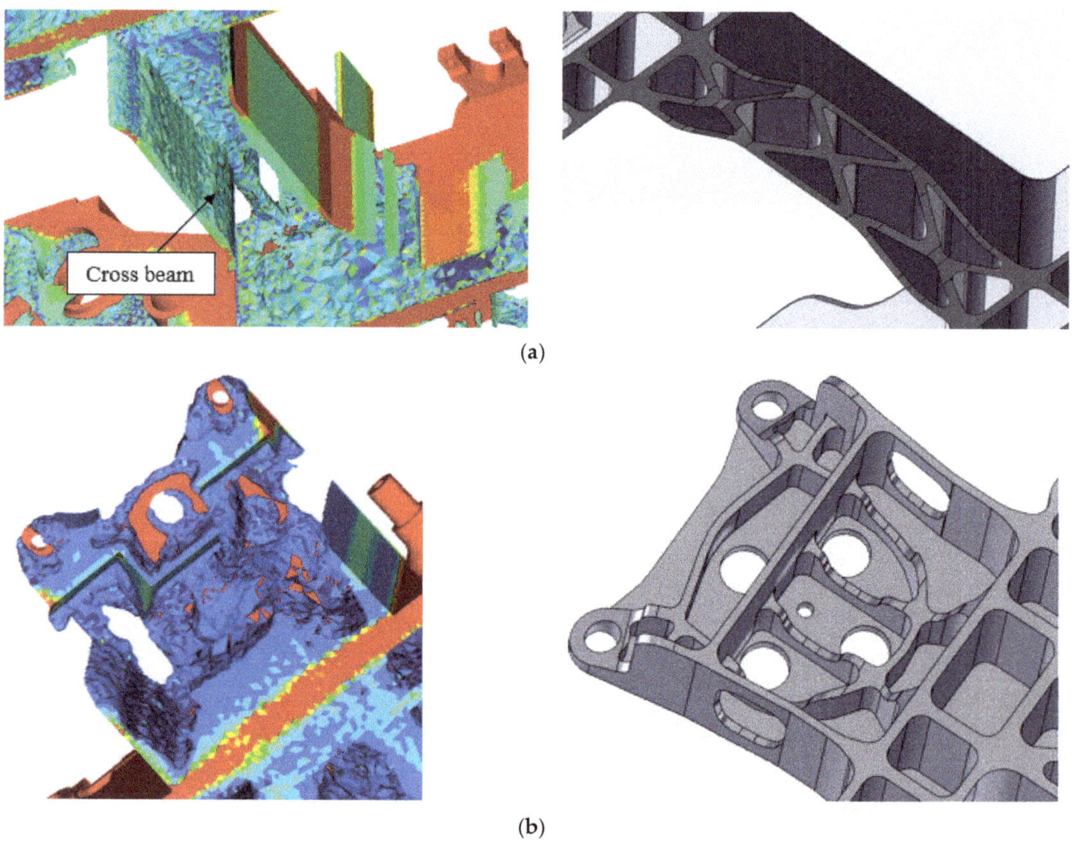

Figure 11. Element density distribution from optimization vs. final bogie frame design: (**a**) cross beam zone, (**b**) secondary suspension zone.

To conclude the activity, the innovative frame was tested both statically and under fatigue conditions in accordance with the reference standard [29], considering all the present load cases. The results were positive, and improvement was observed compared to the original frame, showing overall lower and better-distributed stresses. This once again confirms the effectiveness of the sensitivity analysis process. It allowed for a drastic reduction in computational times to achieve an innovative design for a complex component such as the one studied, ensuring at least the strength of the original frame under all loads considered for the structural verification of the system. Figure 12, in addition to a global view of the innovative design, shows the stress distribution obtained for case PR01, considered the most critical even from the sensitivity analysis. The maximum stress value reduced by 23% compared to the original case and was localized in a small area under two supports, as highlighted in the reference picture. In these cases, a consequent non-linear analysis would be required to investigate potential local material plasticization mechanisms that could be advantageous in terms of resistance. In addition, a general reduction in stress was observed for fatigue load conditions, as reported in Table 3. Utilization factor was calculated as the ratio between absolute principal stress (FEM) and permissible stress.

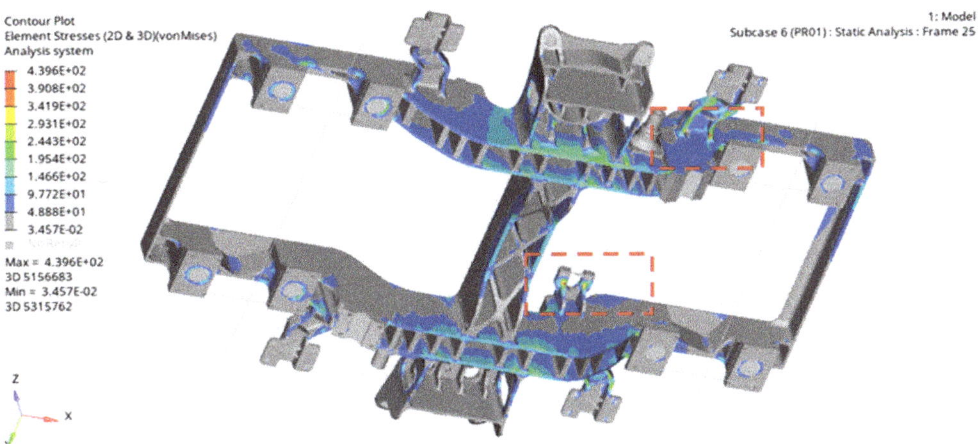

Figure 12. Global view of the innovative design.

Table 3. Fatigue analysis results.

Load Case	Δσ (Absolute Principal Stress) [Mpa]		Δσ (Permissible Stress) [MPa]	Utilization Factor (U) [-]	
-	Original	Redesigned	-	Original	Redesigned
F1-F2	200	162	140	1.43	1.16
PR04-PR05	143	132	140	1.02	0.94
PR08-PR09	107	102	140	0.76	0.73

4. Conclusions and Future Developments

In this article, the research activity was presented, involving the implementation of a complex structural topology optimization applied to a railway bogie frame. This included technological constraints aimed at achieving an innovative design potentially producible through sand-casting. The primary focus of the work centered on conducting a sensitivity analysis to identify the most critical load cases in terms of calculated numerical stress to speed up the design approach. This facilitated the identification of seven reference load cases for a subsequent comparative topology optimization. Positive results were obtained, with a 57% reduction in computational times to achieve a substantially identical design, affirming the validity of the sensitivity analysis. Ultimately, the proposed design resulting from the optimization process fully adhered to the initially imposed technological conditions. The final bogie frame design, featuring innovative shapes and geometries, could potentially facilitate a transition from a welding-based production process to one based on sand-casting, offering potential benefits in terms of time, costs, and mechanical performance. As future developments, accelerating the processes of structural topology optimization could allow them to become an integral part of the design process for new and complex components such as railway structures. Additionally, the potential to integrate mechanical fatigue as an optimization input could provide greater control over crucial outcomes in the field of railway engineering.

Author Contributions: Conceptualization, methodology, software and validation A.C.; review, supervision, project administration E.M. and A.R. All authors have read and agreed to the published version of the manuscript.

Funding: This research was funded by European Union Next GenerationEU (PIANO NA-ZIONALE DI RIPRESA E RESILIENZA (PNRR) MISSIONE 4 COMPONENTE 2, INVESTIMENTO 1.4 D.D. 1033 17/06/2022, CN00000023).

Data Availability Statement: The data presented in this study are available on request from the corresponding author.

Acknowledgments: This study was supported by the MOST Sustainable Mobility National Research Center and received funding from European Union Next GenerationEU (PIANO NAZIONALE DI RIPRESA E RESILIENZA (PNRR) MISSIONE 4 COMPONENTE 2, INVESTIMENTO 1.4 D.D. 1033 17/06/2022, CN00000023). All figures have been provided by Hitachi Rail S.p.A.

Conflicts of Interest: The authors declare no conflicts of interest.

References

1. Billur, S.; Raju, G.U. Mass Optimization of Automotive Radial Arm Using FEA for Modal and Static structural Analysis. *IOP Conf. Ser. Mater. Sci. Eng.* **2021**, *1116*, 012113. [CrossRef]
2. Viqaruddin, M.; Reddy, D.R. Structural optimization of control arm for weight reduction and improved performance. *Mater. Today Proc.* **2017**, *4*, 9230–9236. [CrossRef]
3. Cavazzuti, M.; Baldini, A.; Bertocchi, E.; Costi, D.; Torricelli, E.; Moruzzi, P. High performance automotive chassis design: A topology optimization based approach. *Struct. Multidiscip. Optim.* **2011**, *44*, 45–56. [CrossRef]
4. Pietropaoli, M.; Ahlfeld, R.; Montomoli, F.; Ciani, A.; D'Ercole, M. Design for Additive Manufacturing: Internal Channel Optimization. In Proceedings of the ASME Turbo Expo 2016: Turbomachinery Technical Conference and Exposition. Volume 5B: Heat Transfer, Seoul, Republic of Korea, 13–17 June 2016.
5. Kirsch, K.L.; Thole, K.A. Experimental Investigation of Numerically Optimized Wavy Microchannels Created through Additive Manufacturing. *J. Turbomach.* **2018**, *140*, 021002. [CrossRef]
6. Li, Z.; Zheng, X. Review of design optimization methods for turbomachinery aerodynamics. *Prog. Aerosp. Sci.* **2017**, *93*, 1–23. [CrossRef]
7. Wang, B.; Wang, G.; Tian, K.; Shi, Y.; Zhou, C.; Liu, H.; Xu, S. A preliminary design method for axisymmetric turbomachinery disks based on topology optimization. *Proc. Inst. Mech. Eng. Part C J. Mech. Eng. Sci.* **2022**, *236*, 3313–3322. [CrossRef]
8. Wang, B.; Wang, G.; Tian, K.; Shi, Y.; Zhou, C.; Liu, H.; Xu, S. Topology optimization and laser additive manufacturing in design process of efficiency lightweight aerospace parts. *J. Phys. Conf. Ser.* **2018**, *1015*, 52006.
9. Lang, D.; Radford, D.W. Design Optimization of a Composite Rail Vehicle Anchor Bracket. *Urban Rail Transit* **2021**, *7*, 84–100. [CrossRef]
10. Cho, J.G.; Koo, J.S.; Jung, H.S. A lightweight design approach for an EMU carbody using a material selection method and size optimization. *J. Mech. Sci. Technol.* **2016**, *30*, 673–681. [CrossRef]
11. Koenig, J. Integral consideration of the lightweight design for railway vehicles. In Proceedings of the Young Researchers Seminar, Copenhagen, Denmark, 8–10 June 2011.
12. Miao, B.; Luo, Y.; Peng, Q.; Qiu, Y.; Chen, H.; Yang, Z. Multidisciplinary design optimization of lightweight carbody for fatigue assessment. *Mater. Des.* **2020**, *194*, 108910. [CrossRef]
13. Cascino, A.; Meli, E.; Rindi, A. Dynamic size optimization approach to support railway carbody lightweight design process. *Proc. Inst. Mech. Eng. Part F J. Rail Rapid Transit* **2023**, *237*, 871–881. [CrossRef]
14. Cascino, A.; Meli, E.; Rindi, A. A strategy for lightweight designing of a railway vehicle car body including composite material and dynamic structural optimization. *Railw. Eng. Sci.* **2023**, *31*, 340–350. [CrossRef]
15. Gersborg, A.R.; Andreasen, C.S. An explicit parameterization for casting constraints in gradient driven topology optimization. *Struct. Multidiscip. Optim.* **2011**, *44*, 875–881. [CrossRef]

16. Harzheim, L.; Graf, G. A review of optimization of cast parts using topology optimization: I—Topology optimization without manufacturing constraints. *Struct. Multidiscip. Optim.* **2005**, *30*, 491–497. [CrossRef]
17. Srivastava, P.K.; Shukla, S. Topology Optimization: Weight Reduction of Indian Railway Freight Bogie Side Frame. *Int. J. Mech. Eng.* **2021**, *6*.
18. Yamamoto, M. Non-parametric optimization of railway wheel web shape based on fatigue design criteria. *Int. J. Fatigue* **2020**, *134*, 105463. [CrossRef]
19. Xiu, R.; Spiryagin, M.; Wu, Q.; Yang, S.; Liu, Y. Fatigue life prediction for locomotive bogie frames using virtual prototype technique. *Proc. Inst. Mech. Eng. Part F J. Rail Rapid Transit* **2021**, *235*, 1122–1131. [CrossRef]
20. Guo, F.; Wu, S.; Liu, J.; Zhang, W.; Qin, Q.; Yao, Y. Fatigue life assessment of bogie frames in high-speed railway vehicles considering gear meshing. *Int. J. Fatigue* **2020**, *132*, 105353. [CrossRef]
21. Luo, R.K.; Gabbitas, B.L.; Brickle, B.V. Fatigue Life Evaluation of a Railway Vehicle Bogie Using an Integrated Dynamic Simulation. *Proc. Inst. Mech. Eng. Part F J. Rail Rapid Transit* **1994**, *208*, 123–132. [CrossRef]
22. Lu, Y.; Zheng, H.; Zeng, J.; Chen, T.; Wu, P. Fatigue life reliability evaluation in a high-speed train bogie frame using accelerated life and numerical test. *Reliab. Eng. Syst. Saf.* **2019**, *188*, 221–232. [CrossRef]
23. Luo, R.K.; Gabbitas, B.L.; Brickle, B.V. Dynamic stress analysis of an open-shaped railway bogie frame. *Eng. Fail. Anal.* **1996**, *3*, 53–64. [CrossRef]
24. Mi, C.; Li, W.; Xiao, X.; Jian, H.; Gu, Z.; Berto, F. Lifetime assessment and optimization of a welded a-type frame in a mining truck considering uncertainties of material properties and structural geometry and load. *Appl. Sci.* **2019**, *9*, 918. [CrossRef]
25. Zhou, W.; Zhang, G.; Wang, H.; Peng, C.; Liu, X.; Xiao, H.; Liang, X. Experimental fatigue evaluation of bogie frames on metro trains. *Machines* **2022**, *10*, 1003. [CrossRef]
26. Wang, W.; Bai, J.; Wu, S.; Zheng, J.; Zhou, P. Experimental investigations on the effects of fatigue crack in urban metro welded bogie frame. *Appl. Sci.* **2020**, *10*, 1537. [CrossRef]
27. Lucanin, V.J.; Simic, G.; Milkovic, D.D.; Cupric, N.U.; Golubovic, S.D. Calculated and experimental analysis of cause of the appearance of cracks in the running bogie frame of diesel multiple units of Serbian railways. *Eng. Fail. Anal.* **2010**, *17*, 236–248. [CrossRef]
28. Yüksel, Ç. Microstructural analysis of EN-GJS-450-10 ductile cast iron via vibrational casting. *China Foundry* **2020**, *17*, 272–278. [CrossRef]
29. *EN 13749:2021*; Railway Applications—Wheelsets and Bogies—Method of Specifying the Structural Requirements of Bogie Frames. Available online: https://standards.iteh.ai/catalog/standards/cen/06272177-7cda-4936-a237-125e4631de87/en-13749-2021 (accessed on 23 February 2024).
30. Jameson, A. *Gradient Based Optimization Methods*. MAE Technical Report No (2057). 1995. Available online: http://aero-comlab.stanford.edu/Papers/jameson.gbom.pdf (accessed on 23 February 2024).

vehicles

MDPI

Article

The Individual Drive of a Wheelset and the Problematics of Its Electromechanical Phenomena

Vojtěch Dybala *, Tomáš Fridrichovský and Jan Kalivoda

Department of Automotive, Combustion Engines and Railway Engineering, Faculty of Mechanical Engineering, CTU in Prague, Technická 4, 16000 Prague, Czech Republic; tomas.fridrichovsky@fs.cvut.cz (T.F.); jan.kalivoda@fs.cvut.cz (J.K.)

* Correspondence: vojtech.dybala@fs.cvut.cz

Abstract: This paper deals with the phenomenon of torsion oscillations in railway vehicle drive systems. The main goal is to reduce the risk of presence of torsional oscillations in wheelset drive, eventually to propose systems to effectively identify and eliminate torsional oscillations of wheelsets. Therefore, a simulation wheelset drive model including a detailed model of the asynchronous traction motor control was built. The results of computer simulations show that the torsional oscillations can be effectively eliminated by avoiding the resonance states between the excitation frequencies given by pulse width modulation of the inverter and the eigenfrequencies of the mechanical part of the drive. Furthermore, it was found that the presence of torsional oscillations of the wheels can be detected based on the traction motor current ripple. The wheelset drive model was subsequently implemented in a simulation model of a four-axle locomotive. A new algorithm of an anti-slip protection system that utilizes motor currents was implemented in the model. Simulations show that such an anti-slip protection system can prevent the occurrence of undesired large amplitude of wheelset torsional oscillations. The models and simulation results are presented in detail in the paper.

Keywords: wheelset; inverter; torsion oscillation; pulse-width modulation; electromagnetic torque; railway vehicles; dynamics; MATLAB Simulink; SIMPACK

1. Introduction

In terms of development of traction drives for modern rail vehicles, efforts have generally been aimed at more efficient and powerful systems. Increasing the power up to theoretical physical limits (meaning especially adhesion limits) also reveals new negative physical phenomena. Torsional oscillations of traction drive components, which are discussed in this paper, play an important role among these undesirable effects.

Recently, problems with Deutsche Bahn locomotives class 145 (Bombardier TRAXX family) have surfaced. The relative rotation of the wheel disc with respect to the wheelset axle has been observed within locomotive maintenance (see Figure 1) [1]. Such slip indicates failure of the press-fitted joint that holds the wheel and the axle together. This means the friction between two parts is reduced enough to allow the wheel to freely rotate or shift along the axis for a short time.

This condition can be dangerous, especially in situations when the vehicle passes through a curve or railroad switch. At that moment, significant transversal forces which act in contact with a wheel flange and rail head [2] may cause a shift in the wheel along the axis, thus pushing the wheels closer to each other. This leads to a decrease in the lateral distance between wheels, e.g., back-to-back dimension, as schematically presented in Figure 2. If the distance between wheels decreases under limit values, the vehicle can derail.

As the most probable reason for such shifts, the phenomenon of torsion oscillations was assumed. An effort to investigate the occurrence of torsional oscillations motivated the research in the area of traction drives of modern locomotives and other high-power rail vehicles. Mathematical simulation models are usually used as the principal research tool. Complex simulation models of a whole railway vehicle were presented by Markovic et al. [3] and Trimpe et al. [4]. The research in the field of torsion oscillations in traction drives of railway vehicles focuses on the development of control systems, which are used for an elimination of undesired behaviour of traction drives during the vehicle operation, for example, anti-slip control systems [5]. However, many other control systems can be utilized in modern traction drive systems. Efforts in this research area include controllers and control systems for solving dynamic issues with an adhesion loss and re-gain [6,7], control systems for an active reduction of vibrations and noise [8–10], predictive slip control systems [11] or systems applying advantages of fuzzy logic for anti-slip control against classic Proportional-Integral-Derivative (PID) controllers [12]. Another area of research of torsion oscillations of traction drives focuses on the investigation of dynamic loading of mechanical parts of drive trains of railway vehicles.

Figure 1. Relative motion of the wheel and the axle. See the mismatch in yellow mark [1].

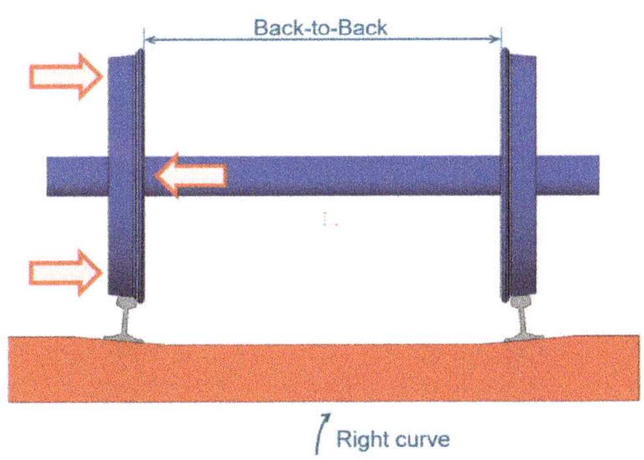

Figure 2. Schematic representation how wheel distance can decrease due to one-side wheel shift. In this case, the loss of connection occurs for the left wheel.

This research focuses especially on dynamic loading of wheelsets and its relationship to electrical parameters of the drive train. In [13], self-excited torsion oscillations of the wheelset were analysed. The same problematics from the perspective of an adhesion loss were analysed by Liu et al. [14], utilizing a simulation model of a wheelset drive with two degrees of freedom. Qi and Dai in [15] focused on the electro-mechanical coupling between harmonic components of the traction motor torque and the wheel wear, whereas in [16], the influence of traction dynamics on the wheel polygonization was analysed. The negative impacts of electro-mechanical coupling in the asynchronous motor drive systems were investigated by Winterling [17], who examined complex dynamic behaviour of drive systems of locomotives. Szolc et al. [18] focused on dynamic electro-mechanical coupling effects in machine systems driven by asynchronous motors. The system analysis presented by Schneider [19,20] describes the effects of a wheelset overloading caused by the combined load from torsion oscillations and acting lateral forces. The importance of the research of the harmonic components was also pointed out by Takahashi et al. [21], who focused specifically on the efficiency of traction motors for railway vehicles, or in the work of Lu et al. [22], which proves, that pulsating motor torque has significant influence on the wheel-rail contact forces and longitudinal and vertical vibrations within the vehicle bogie.

The work presented in this paper utilizes a complex model of the drive system of a locomotive. Unlike the works published so far, special emphasis is placed on two innovative ideas: elimination of torsion system resonances via torsional stiffness optimization and improving the sensitivity of the traction motor control system to the detection of presence of torsional oscillations using observation of traction motor phase currents.

2. Object of Interest

The parameters of the mathematical models in this study are based on the design of modern four-axle locomotives. These locomotives are equipped with a pair of traction bogies, both with two individually powered wheelsets. The asynchronous traction drives provide overall power output between 6 and 7 MW. Typical representatives of this category of locomotives are Siemens ES64U4, Siemens Vectron, or Skoda 109E (Figure 3).

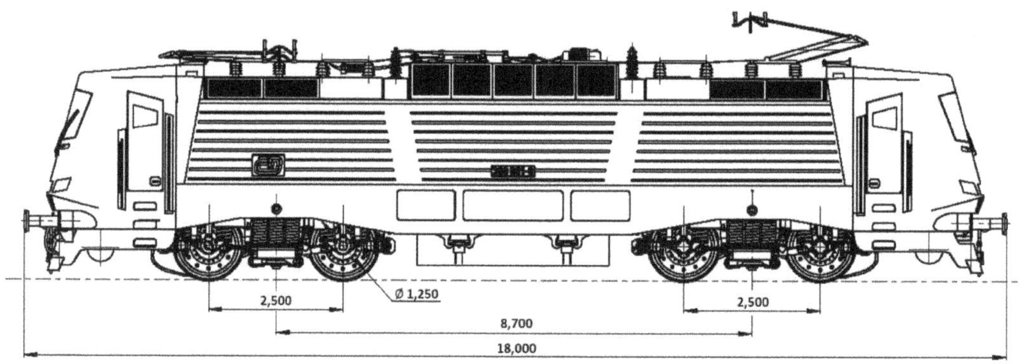

Figure 3. Modern asynchronous locomotive: Skoda 109E (preliminary design) [23].

Mathematical modelling was realized in two steps. In the first step, simulations were performed with a wheelset drive model containing a torsional-flexible system of a single wheelset and its electrical drive. In the second step, a model of a whole vehicle was utilized. The vehicle model uses the wheelset drive models and moreover includes a detailed model of the mechanical system of the whole vehicle with implemented real external conditions such as adhesion, drive resistances, and track irregularities. These two mathematical models, namely the wheelset drive model and vehicle model, respectively, are described in the sections below.

2.1. Simulation Model of the Wheelset Drive

A complete drive of any railway vehicle is an extensive system. The model of the wheelset traction drive is split into four main groups which define four main sub-models—a mechanical part, an electrical-control part, a wheel-rail contact and a train drive dynamics model.

The connection of these four sub-models creates the whole wheelset model. A scheme representing sub-models and their relations (outputs/inputs) is shown in Figure 4. This model of a single drive was created within Matlab Simulink R2017a software.

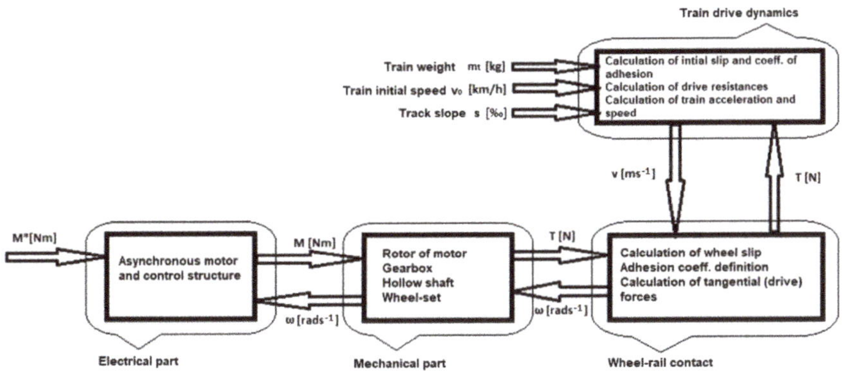

Figure 4. Wheelset model—scheme of the structure with main parts [24].

2.1.1. Mechanical Part

A fully suspended hollow shaft drive (see Figure 5) is addressed in this study. Such a design allows decreasing unsprung masses and forces resulting from track irregularities, which is important for higher velocities of railway vehicles within tracks of common quality. The mechanical part of the wheelset drive consists of these main parts:

- rotor of the traction motor,
- gearbox with a pinion and a gearwheel,
- hollow shaft with elastic articulations/clutch/coupling,
- wheelset consisting of an axle and wheels.

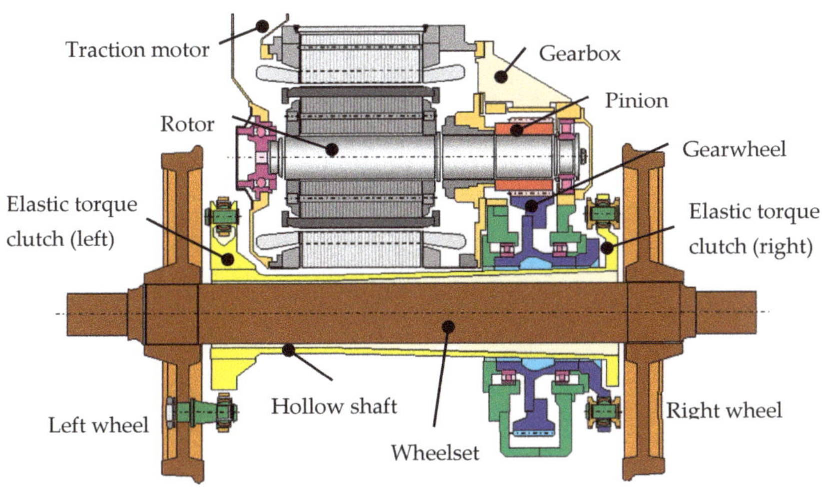

Figure 5. Cross-section of a modern, fully suspended drive of a locomotive [25].

The fully suspended wheelset drive is schematically presented in Figure 6. For the purpose of mathematical description, the mechanical part of wheelset drive was divided in to seven rotational masses mutually connected by torsion springs designated as *kt*. Thus, it forms a torsion system with seven degrees of freedom.

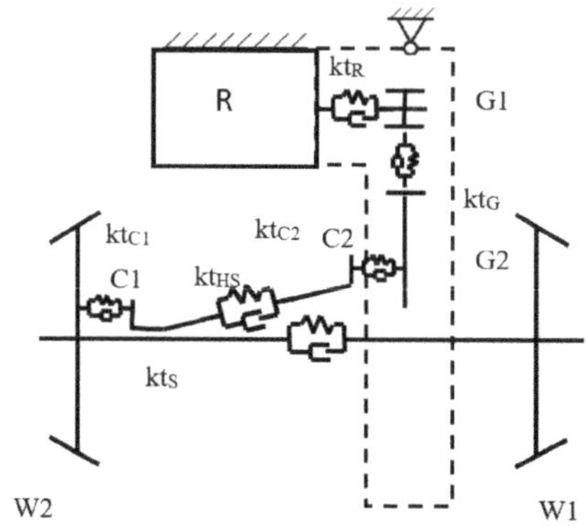

Figure 6. Schematic representation of a fully suspended drive - R—rotor, G1—pinion, G2—gear wheel, C2—right coupling, C1—left coupling, W2—left wheel, W1—right wheel.

2.1.2. Electrical Part—Generation and Control of Torque

The electrical part is focused on the traction motor and its control. An asynchronous motor is considered. The mathematical description corresponds to voltage-flux Equations (1)–(8):

$$u_{1\alpha}(t) = R_1 i_{1\alpha}(t) + \frac{d\Psi_{1\alpha}(t)}{dt} \tag{1}$$

$$u_{1\beta}(t) = R_1 i_{1\beta}(t) + \frac{d\Psi_{1\beta}(t)}{dt} \tag{2}$$

$$u_{2\alpha}(t) = R_2 i_{2\alpha}(t) + \frac{d\Psi_{2\alpha}(t)}{dt} + p_p \omega_m(t)\Psi_{2\beta}(t) = 0 \tag{3}$$

$$u_{2\beta}(t) = R_2 i_{2\beta}(t) + \frac{d\Psi_{2\beta}(t)}{dt} - p_p \omega_m(t)\Psi_{2\alpha}(t) = 0 \tag{4}$$

where u_1 is stator voltage, i_1 is stator current, Ψ_1 is stator flux, R_1 is stator ohmic resistance, u_2 is rotor voltage, i_2 is rotor current, Ψ_2 is rotor flux, R_2 is rotor ohmic resistance, p_p is number of pole pairs, and ω_m is rotor mechanical angular speed. Voltage equations, particularly (3) and (4), contain angular speed of the rotor, ω_m, which means there is a relation between electrical and mechanical values in the motor.

Stator and rotor flux are given by currents and inductances, as defined in Equations (5)–(8):

$$\Psi_{1\alpha}(t) = L_1 i_{1\alpha}(t) + L_h i_{2\alpha}(t) \tag{5}$$

$$\Psi_{1\beta}(t) = L_1 i_{1\beta}(t) + L_h i_{2\beta}(t) \tag{6}$$

$$\Psi_{2\alpha}(t) = L_2 i_{2\alpha}(t) + L_h i_{1\alpha}(t) \tag{7}$$

$$\Psi_{2\beta}(t) = L_2 i_{2\beta}(t) + L_h i_{1\beta}(t) \tag{8}$$

where L_1 is stator inductance, L_2 is rotor inductance, and L_h is coupled inductance.

Resulting torque Equation (9) is derived from the previous Equations (1)–(8):

$$M(t) = \frac{3}{2} p_p \left(i_{1\beta}(t)\Psi_{1\alpha}(t) - i_{1\alpha}(t)\Psi_{1\beta}(t) \right) \tag{9}$$

where M is the motor torque

The input for the motor control is the required torque of the motor. The output from the motor control is the mechanical torque of the motor. The control is based on the vector control scheme which is shown in Figure 7. Such an approach allows us to control alternate current machine with direct current values (currents) similar to direct current motors, where torque is proportional to the direct current within a certain range of rotor speed.

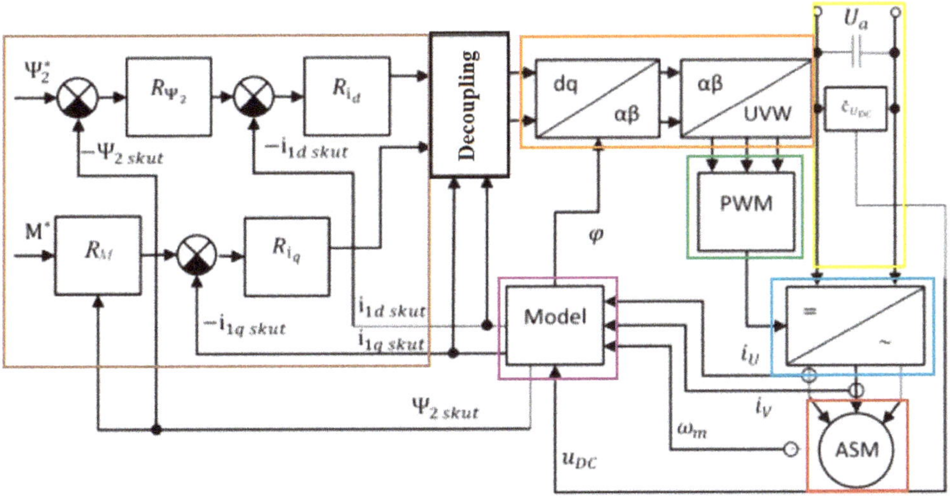

Figure 7. Vector control scheme [26].

The required value of torque is determined from the actual requirement for velocity of the vehicle, which is also controlled via an anti-slip protection block. This block permanently checks angular velocities and angular accelerations of all wheelsets (in fact, rotors) and via mutual comparison determines the occurrence of excessive slippage or loss of adhesion. This control is included in the controller's block of the electrical model (see brown rectangle in Figure 7). The other parts of the electrical model in Figure 7 are: decoupling block (black rectangle), coordinate systems transformation block (orange rectangle), pulse-width modulation (PWM) block (green rectangle), supply voltage block (yellow rectangle), inverter block (blue rectangle), asynchronous motor (ASM) block (red rectangle) and ASM model block (violet rectangle).

2.1.3. Wheel-Rail Contact—Adhesion Model

The maximum tangential force transmissible in a wheel-rail contact T is given by Equation (10):

$$T = \mu \cdot Q \tag{10}$$

where μ is the adhesion coefficient and Q is the vertical force, i.e., the wheel force.

The calculation of the adhesion coefficient μ is based on the adhesion theory of Polach [27], considering minor updates allowing wider variability for the setting of friction parameters. The resulting adhesion coefficient μ applied within the wheelset drive model in dependence on the relative slip and vehicle velocity is presented in Figure 8.

The longitudinal slip s_x is defined as a division of slip speed and vehicle speed according to Equation (11), where r_k is the rolling radius of the wheel, ω_k is the angular speed of the wheel, and v is the velocity of the vehicle:

$$s_x = \frac{r_k \cdot \omega_k - v}{v} \tag{11}$$

The range of velocities of 5–20 m·s^{-1} corresponds to the range of running velocities of the vehicle at which the traction motors are able to develop maximum torque. At higher running speeds, the torque of the traction motors is limited by the maximum power output.

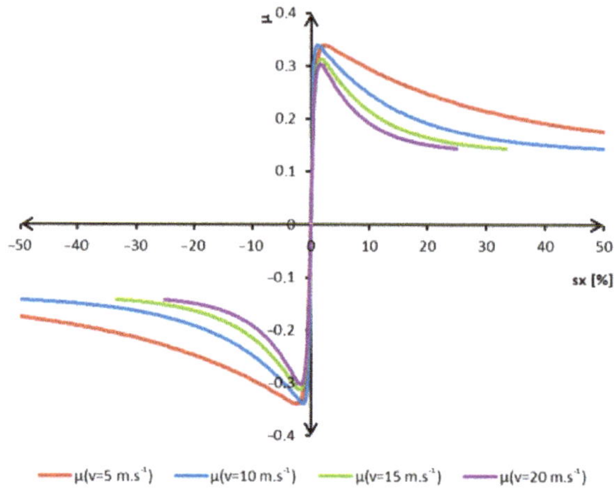

Figure 8. Friction characteristics showing the relation between the slip and the friction coefficient for different velocities of a vehicle.

2.1.4. Train Drive Dynamics—Velocity Calculation

This submodel calculates analytical formulas from the traction mechanics to provide the velocity of a simulated vehicle run. The calculations are based on the balance between accelerating forces and resistance forces. The difference between these forces determines the resulting longitudinal force acting on the vehicle, which in the relation to the mass of the vehicle results in acceleration, a. The final product of these calculations is the speed of the modelled train, v (Equation (12)), which is the result of the time integration of the initial speed v_0 and acceleration a.

$$v = v_0 + \int_0^t a \, dt \tag{12}$$

2.2. Simulation Model of the Vehicle

Using the wheelset drive model, the vehicle model was built (Figure 9). The vehicle model contains all important components reflecting the drive scheme in Figure 5 and other vehicle features such as suspension, bogie frame, carbody etc. The mechanical part of the vehicle model was built within SIMPACK 2019 software. This software also considered an adhesion model and other outer conditions (drive resists, track irregularities, track curvatures etc.). Electrical and control parts of the model were taken from the wheelset model.

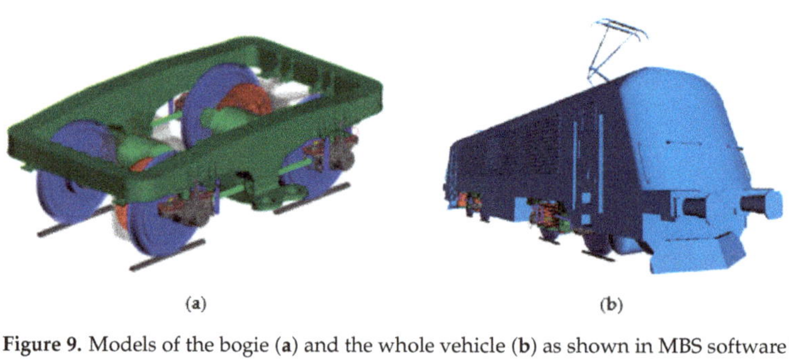

Figure 9. Models of the bogie (**a**) and the whole vehicle (**b**) as shown in MBS software SIMPACK.

3. Simulation Results

3.1. Electromagnetic Excitation and Torsion Vibrations

For purposes of the research explained in this section, the wheelset model of the single drive was utilized. The fundamental behaviour of the three-phase asynchronous traction motor supplied from the inverter could be explained on its one winding phase. The phase current is rippled. The essence of the traction motor current ripple Δi_d (see Figure 10 and Equations (13)–(17)), is described in the literature, e.g., [28].

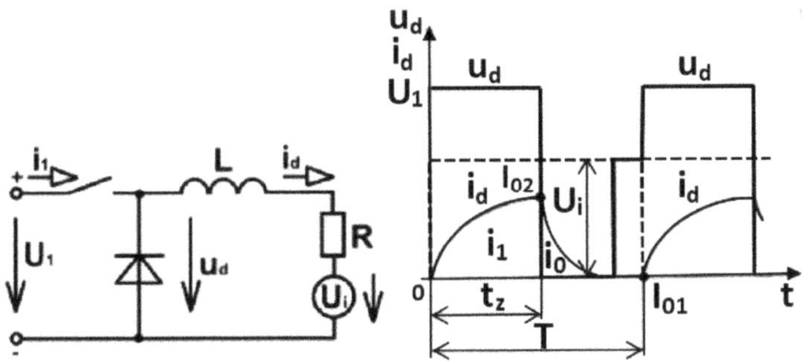

Figure 10. Pulse inverter scheme and the course of the output voltage and the current.

When the switch is ON, the supply voltage U_1 is put on the L-R-U_i load, and in the circuit there is an immediate direct voltage u_d and direct current i_d increase according to Formulas (13) and (14):

$$R \cdot i_d + L \cdot \frac{di_d}{dt} + U_i = U_1 \tag{13}$$

$$i_d = I_{01} \cdot e^{-\frac{t}{\tau_d}} + \frac{U_1 - U_i}{R} \cdot \left(1 - e^{-\frac{t}{\tau_d}}\right) \tag{14}$$

where R is ohmic resistance, L is inductance, t is time, dt is a time derivative, I_{01} is the initial value of direct current, I_{02} is the final value of direct current, and τ_d is the machine time constant. The induced voltage U_i is induced in the load as one of its characteristics.

When the switch is OFF there is no supply voltage U_1 on the load. In the circuit, there is no immediate direct voltage u_d, and the direct current i_d decreases according to Equations (15)–(17). Only the induced voltage U_i in the load is present:

$$R \cdot i_d + L \cdot \frac{di_d}{dt} + U_i = 0 \tag{15}$$

$$i_d = I_{02} \cdot e^{-\frac{t}{\tau_d}} - \frac{U_i}{R} \cdot \left(1 - e^{-\frac{t}{\tau_d}}\right) \tag{16}$$

$$\Delta i_d = I_{02} - I_{01} \tag{17}$$

A real asynchronous motor supplied from an inverter has a rippled current by many harmonics, including the switching frequency and its multiples, the first harmonic of the supplying voltage and its multiples, frequency sidebands etc. Harmonic components are the source of the excitation of torsion vibrations in the mechanical part of the traction drive. More specifically, the current ripple causes the ripple of the electromagnetic torque. The excitation frequencies of the inverter were determined in accordance with [29] by Equations (18) and (19):

- For odd multiples of the switching frequency f_{PWM}:

$$f_i = k \cdot f_{PWM} \pm 2 \cdot l \cdot f_1 \tag{18}$$

where: f_1 is the first harmonic of the supply voltage, (i = 1, 2, 3, . . .); (k = 1, 3, 5, . . .); (l = 1, 2, 3, 4, . . .);

- For even multiples of the switching frequency f_{PWM}:

$$f_i = m \cdot f_{PWM} \pm (2 \cdot n + 1) \cdot f_1 \tag{19}$$

where: (i = 1, 2, 3, . . .); (m = 2, 4, 6, . . .); (n = 0, 1, 2, 3, . . .)

The basic simulation and modal analysis of the torsion system provide essential information about eigenfrequencies, content of harmonics (excitation frequencies involved in the electromagnetic torque of the traction motor) and their resulting potential resonance states. This complex knowledge is presented via Campbell's diagram (Figure 11), which shows six eigenfrequencies of the torsion system (from the second the seventh, as the first eigenvalue corresponds to the natural rotation of the torsion system), ten excitation frequencies (from the third multiple of the first harmonics f_1 through multiples of switching frequencies and their side bands) and five potential resonances (from I to V). Further simulations were focused on the pinion and its respective resonance states:

- III—resonance of the fifth eigenfrequency with the frequency sideband of the switching frequency f_{PWM}.
- V—resonance of the seventh eigenfrequency with the frequency sideband of the third multiple of the switching frequency f_{PWM}.

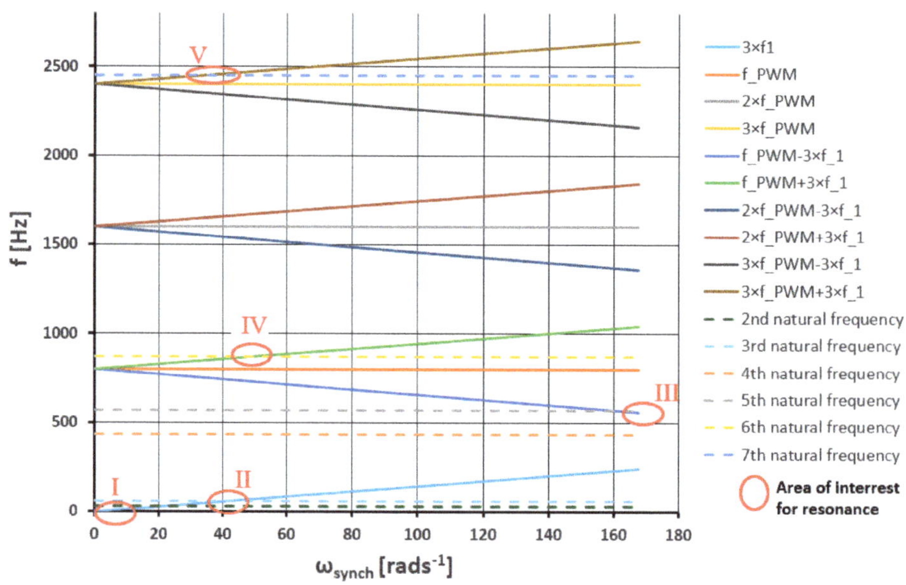

Figure 11. Campbell diagram.

To study these resonance states, simulations of the acceleration of the vehicle from 0 kmh^{-1} up to the speed of 86 kmh^{-1} (approx. 23.8 ms^{-1}, $\omega_R = \omega_P \approx 165$ rads^{-1}) were performed. During the simulations, the following limitations of traction drive were applied: the maximal change in the traction motor drive torque over time was 1000 Nms^{-1}, with maximal traction motor torque of 8000 Nm in the whole speed range. Figure 12 presents the course of the pinion driving torque within this simulation with an excited resonance in the time around 27 s.

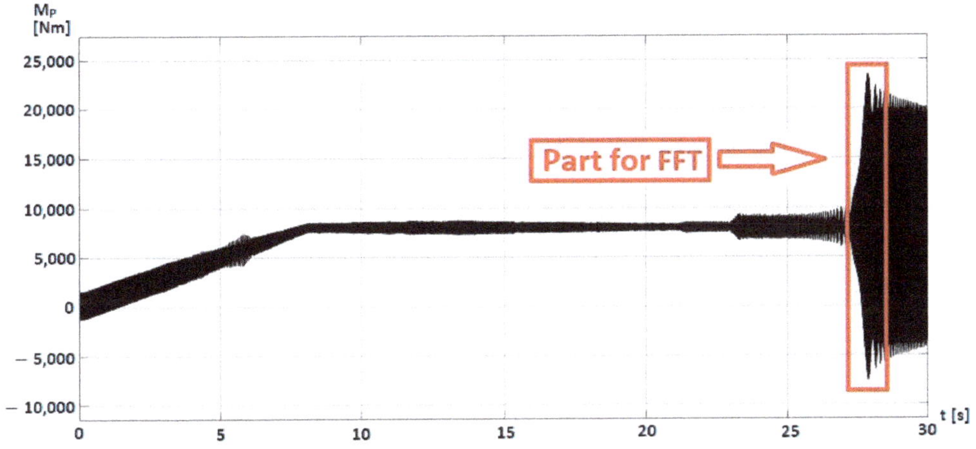

Figure 12. Driving torque of the pinion.

The area of resonance formation marked by the red rectangle in Figure 12 was analysed by Fast Fourier Transformation (FFT) analysis (Figure 13).

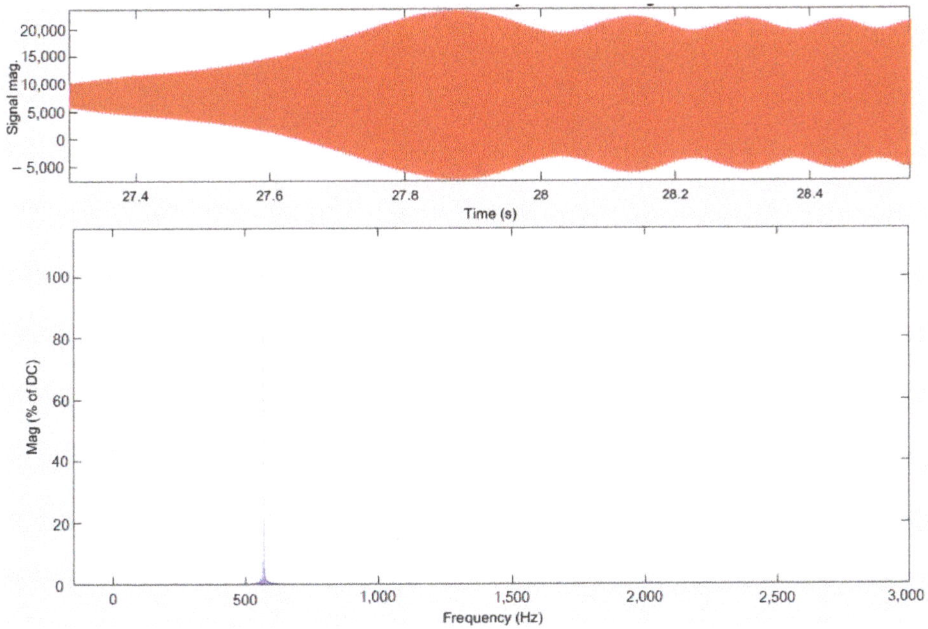

Figure 13. Driving torque of the pinion—FFT analysis.

During the drive through the speed spectrum, very significant torsion vibrations were excited. The FFT analysis revealed that these enormous vibrations have a frequency of 568 Hz, which is close to the fifth eigenfrequency of the torsion system. The excitation frequency of the sideband is 570 Hz, which provideds evidence of excitation of the supposed resonance state III.

To investigate the harmonics with respect to speed, simulations were carried out for vehicle speeds of 10 kmh^{-1}, 25 kmh^{-1}, 52 kmh^{-1}, 75 kmh^{-1} and 86 kmh^{-1} (Figure 14). The figure shows increasing magnitudes, which were expected. The most dominating component is the third multiple of f_1.

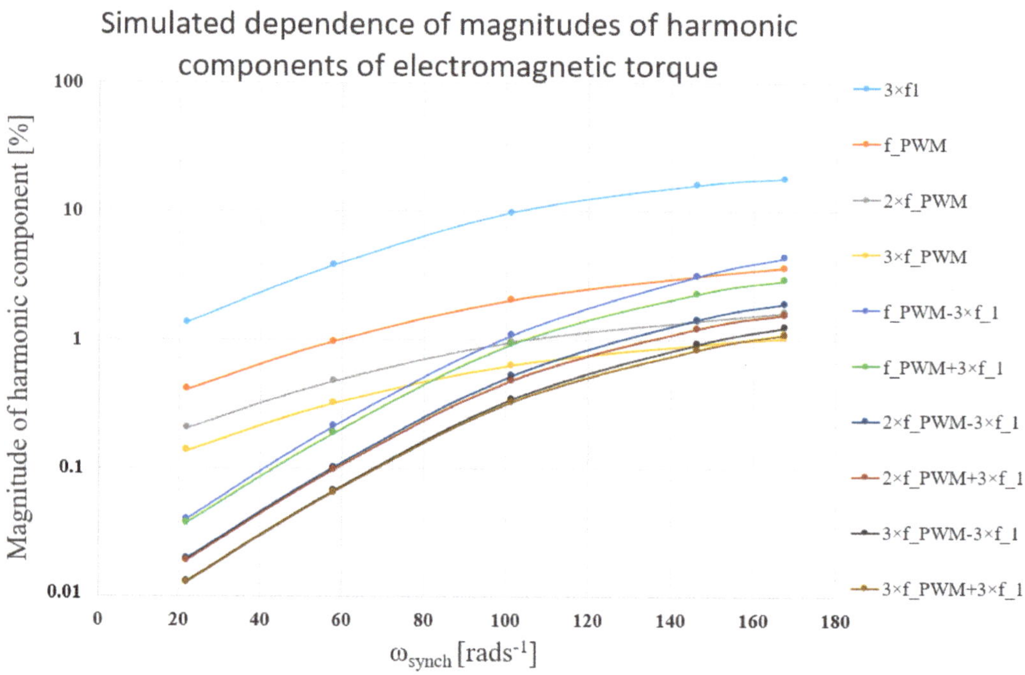

Figure 14. Harmonic components magnitude function [30].

In the context of Figure 14, the first way of elimination is to optimize the fifth eigenfrequency via change in torsion stiffness k_{tR}—which represents stiffness of the shaft between the rotor and pinion. The excitation magnitudes decrease with decreasing speed, and therefore, the potential of harmonic components which could excite a resonance state decreases as well. This leads to the increase in k_{tR} and the eigenfrequency in this case. The resonance state excitation can also be eliminated by moving the specific eigenfrequency into lower values, i.e., out of the excitation frequency range.

The set of simulations for modified k_{tR} and respective fifth eigenfrequency was carried out as presented in Figure 15. The stiffness ratio is defined towards the nominal value applied in the first simulation (Figure 13).

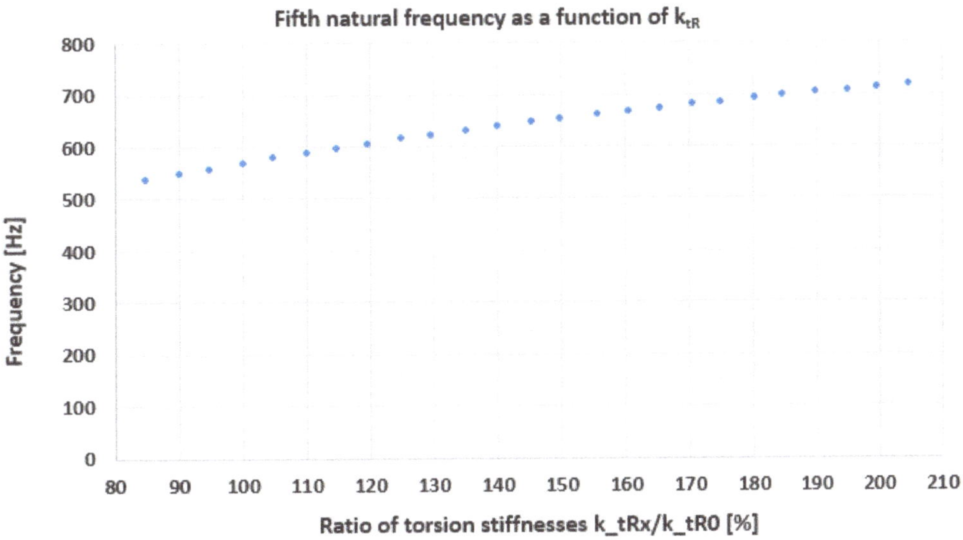

Figure 15. Set of torsion stiffnesses.

The approach to eliminate the resonance for the fifth eigenfrequency via moving it out of the range of excitation frequencies is supported by the Campbell diagram, which indicates that the fifth eigenfrequency should be under 560 Hz. As the asynchronous motor works with a slip, the stiffness was set up for k_{tR1} up to 85%.

The second way of elimination of resonance states was tested via a set of simulations for the torsion stiffnesses from 105% to 205%. It was assumed to reduce the potential of excitation harmonic components by increasing the eigenfrequencies. Figure 16 shows the result of this set of simulations and the fact that the excited resonance state of the pinion loading torque was significantly reduced by that approach. Equivalent results were reached by the reduction in torsion stiffness (from 85% to 90%). The Figure 17 summarizes results of FFTs from these simulations with respect to maximum magnitudes of excited resonance states.

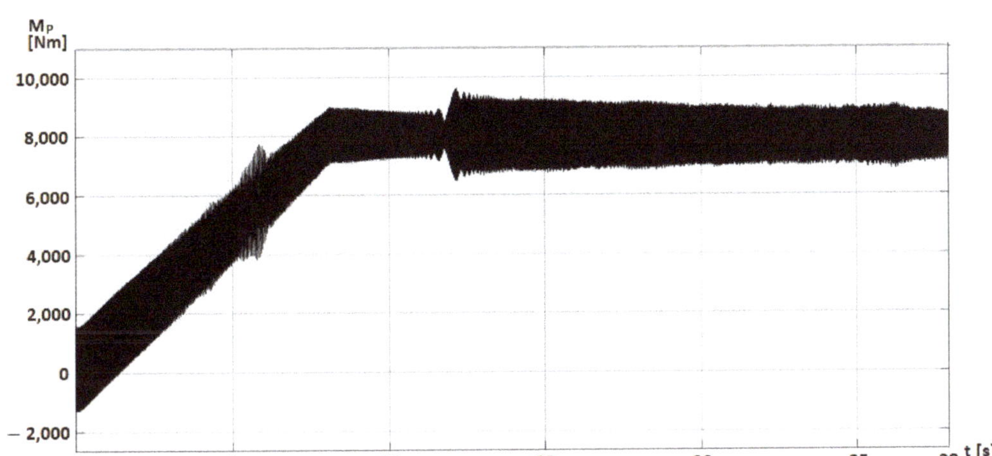

Figure 16. Driving torque of the pinion for increased torsion stiffness k_{tR} up to 205%.

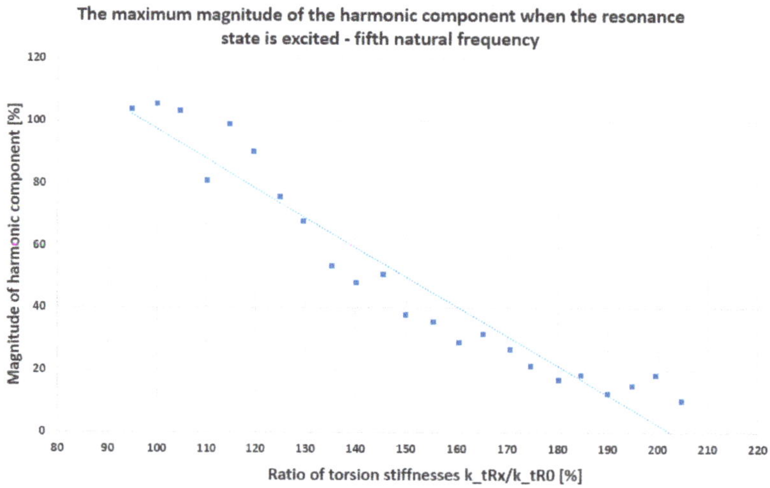

Figure 17. Magnitudes of harmonics as a function of the torsion stiffness k_{tR}—the resonance state of the fifth eigenfrequency.

It is clear that with increasing of the stiffness k_{tR}, the magnitude of the studied harmonic component decreases, as well as when the eigenfrequency is moved out of the range of excitation frequencies via decreasing of the stiffness k_{tR}. This is the conclusion which confirms the initial assumption for carrying out the simulations.

3.2. Phenomena Occurring during Loss of Adhesion

Another situation with a high potential for torsional oscillations is running a wheelset on the limit of adhesion and sudden decrease in friction between wheel and rail. Two cases of such a situation can be distinguished:

a. the wheelset loses its adhesion instantly (prestressed components such as an axle may quickly release its torsion energy);
b. the wheelset loses its adhesion gradually (energy accumulated in the wheelset is gradually released).

The first situation usually leads to rapid acceleration of the wheelset (likewise the rotor of traction motor) and relatively quick reaction of the anti-slip protection. The second situation is considerably harder to recognize. The axle usually starts to oscillate on one end, but the other wheel still holds the adhesion, so the wheelset as a whole does not accelerate to higher speeds. Absence of rapid change then results in undesirable delays in the reaction time of the anti-slip protection system. Such a situation is shown in Figure 18. At time 15 s, the adhesion of the right wheel W1 (see Figure 7) decreases from 0.4 to 0.2, and the wheelset subsequently starts to oscillate within maximum torsion amplitudes up to 0.02 rad (corresponding to approximately 1 degree). The anti-slip protection system recognizes the non-standard behaviour of the drive around time 17 s and decreases engine torque, which leads to gradual disappearance of wheelset oscillations.

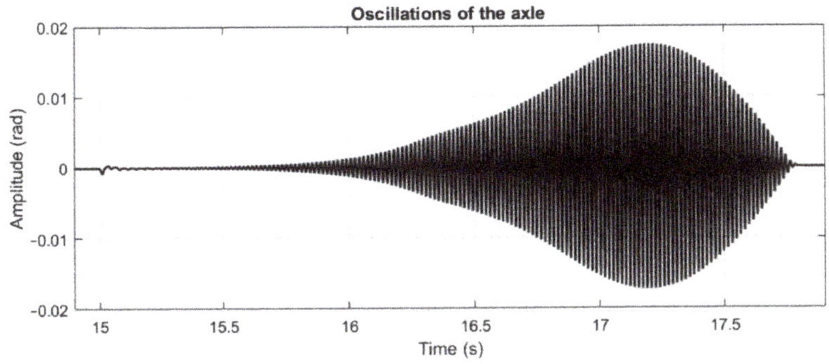

Figure 18. Time course of wheelset torsion amplitudes. Oscillations are performed via high pass filter.

Such oscillations are observable via phase currents of the asynchronous motor. Figure 19 shows time course of phase current i_{1q}, both as an absolute value (solid line) and as a filtered value (dashed line). It is noticeable that the current starts to ripple significantly between the time of 16 and 16.5 s. The torsional oscillations of the wheelset increase undisturbed up to the time of 16.8 s, when anti-slip protection reacts and decreases torque of the motor, which also causes the disappearance of the oscillations.

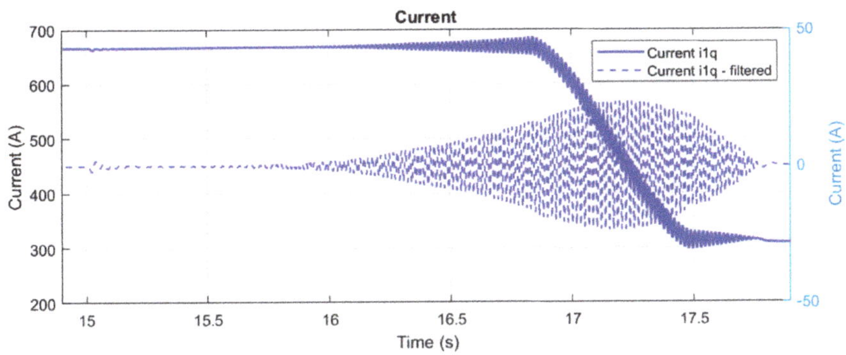

Figure 19. Time course of control current in the motor during torsional oscillations. The decreasing curve corresponds to the reduction in torque. Filtered oscillations are performed via high pass filter.

The important fact is a correlation between frequencies of wheelset oscillations and oscillations of phase current i_{1q}, which is the result of the presence of angular velocity in the voltage equation of asynchronous motors (4). Frequencies are identical as can be seen in Figure 20, which means that the wheelset oscillations can be identified via observation of electrical values. This allows us to check actual state of the wheelset (whether it is oscillating or not).

This approach allows us to implement an additional block into the control scheme of the drive that monitors the phase currents of the motor in order to detect the presence of oscillations in the wheelset. This block is able to recognize the wheelset oscillations earlier then a standard anti-slip protection system.

A set of simulations was performed in order to prove the relation between mechanical and electrical oscillations in the drive. The simulations were focused on the situation of the loss of adhesion on the left or right wheel considering different velocities of the vehicle and different amounts of motor torque. The range of motor torques was between 60 and 100% of the maximum value.

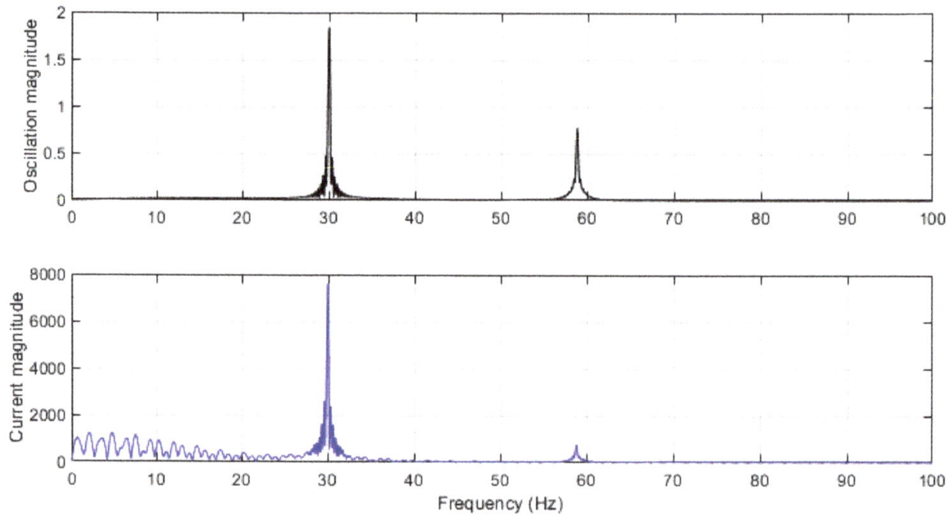

Figure 20. Power spectral density of simultaneous oscillation of wheelset and control current i_{1q}.

Comparison of mechanical and electrical oscillations allows us to determine the relation between them, which is shown in Figure 21. It can be seen that the relation between the amplitudes of the motor current and the wheelset oscillations is almost linear. This applies to slippage on both (left and right) wheels. Using these linear dependencies, it is possible to assess whether the wheelset oscillations detected by the oscillations of the motor currents are harmful or not.

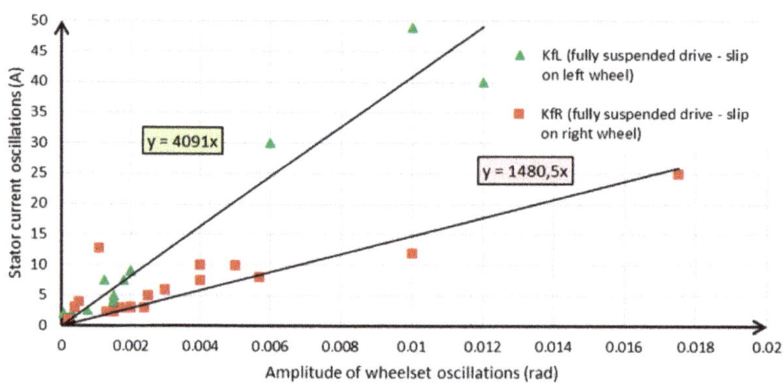

Figure 21. Relation between amplitudes of oscillations of the axle and currents for the partially and fully suspended drive with relation to slip of a particular wheel [31].

3.3. Simulations of Vehicle Model

The wheelset model was utilized in the vehicle model. The control structure containing standard anti-slip protection shown in Figure 4 was upgraded with a part measuring and evaluating values of motor phase currents. This was done in order to recognize the appearance or evolution of mechanical oscillations into higher values. These oscillations cause enormous load to the axle and other components of the drive, but the anti-slip protection does not detect them. The system is able to recognize a situation when the wheelset has completely lost adhesion and starts to accelerate together with the rest of the drive, but not the situation when the wheelset oscillates. This is due to fact the anti-slip protection compares angular speeds and angular accelerations of all rotors. When one of the rotors rotates much faster than the others, the control block evaluates this as a slippage and reduces torque of the motor. A similar situation occurs when one of the rotors starts to accelerate faster than the others, or when the acceleration exceeds physically real values.

Regardless of whether the wheelset starts to accelerate or not, oscillations of one wheel always transfer through the drivetrain into the rotor and influence electrical values of the motor itself. Observing the motor current may help recognize situations when the wheelset oscillates. This may be a source of the information for a quicker reaction of the anti-slip protection and a reduction in oscillations.

A comparison between standard and updated controls is shown in Figure 22. This figure shows time courses of original and updated settings, considering the same bounding conditions of a vehicle running through the curve. The wheelset starts to oscillate between 15.5 s and 17.75 s. Additional information about the presence of oscillations allows decreasing the torque faster by approximately 0.2 s, so oscillations amplitudes cannot evolve to higher values. Therefore, it is possible to reduce significantly high amplitudes of oscillations that occur in the wheelset—approximately to 20% of the original value, similarly to time duration of the phenomenon.

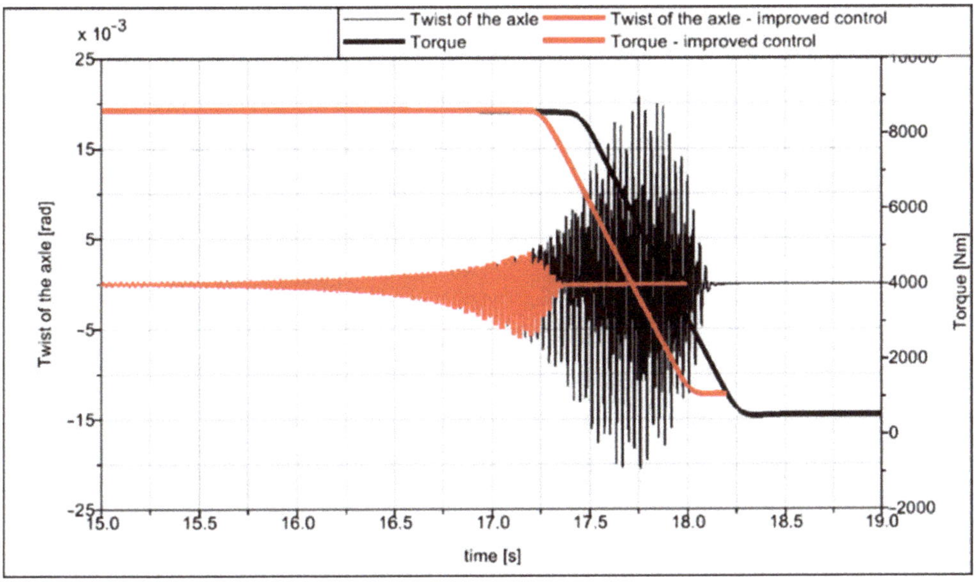

Figure 22. Comparison of different time courses—standard torque control and the improved torque control that assumes currents in the motor.

The important aspect is that such an upgrade of drive control is mainly a software task without the necessity for any design changes in mechanical and power parts of the wheelset drive. Therefore, this approach may be relatively easily implemented on already running vehicles that are operated frequently and thus spare significant costs connected with the presence of torsion oscillations.

4. Conclusions

Wheelset torsional oscillations are an undesirable and potentially very dangerous phenomenon. Present anti-slip systems are not capable of effectively detecting and eliminating such torsional oscillations.

Computer simulations realized with the wheelset drive simulation model revealed that harmonic components of the electromagnetic torque of the asynchronous traction motor can excite significant torsion vibrations, which cause torsion overloading of the mechanical components. Such problems can be eliminated by optimizing the eigenfrequencies of mechanical parts of the wheelset drive by modifying torsion stiffnesses of coupling elements. Furthermore, it was found that the presence of torsional oscillations of the wheels can be detected based on the traction motor current ripple. This finding was used in the design of a new algorithm of anti-slip protection system. This algorithm was implemented on a four-axle locomotive model.

The simulations proved that wheelset torsional oscillations can be observed in advance within currents of the asynchronous traction motor. Our results showed that the improved anti-slip protection, implementing identification of torsion vibrations via motor currents, reacts faster and enables preventing the loss of traction and development of large amplitudes of torsion oscillations jeopardizing a wheelset. This new algorithm for anti-slip protection can also be implemented in software of vehicles that are already in operation and for which any changes in their mechanical design are very difficult, if not impossible.

Author Contributions: Conceptualization, T.F., J.K. and V.D.; methodology, T.F. and V.D.; software, T.F. and V.D.; validation, T.F. and V.D.; formal analysis, T.F. and V.D.; investigation, T.F. and V.D.; resources, T.F. and V.D.; data curation, T.F., J.K. and V.D.; writing—original draft preparation, T.F., J.K. and V.D.; writing—review and editing, T.F., J.K. and V.D.; visualization, T.F., J.K. and V.D.; supervision, J.K.; project administration, T.F., J.K. and V.D.; funding acquisition, T.F., J.K. and V.D. All authors have read and agreed to the published version of the manuscript.

Funding: This research was funded by The Technology Agency of the Czech Republic, programme National Competence Centres, project # TN01000026 Josef Bozek National Center of Competence for Surface Transport Vehicles and by The Czech Technical University and a related grant (No. SGS20/120/OHK2/2T/12).

Data Availability Statement: Data are contained within the article.

Acknowledgments: Support from the tutor of both authors, Doc. Ing. Josef Kolář, CSc., is gratefully acknowledged for both theoretical and practical components of the research. Support during laboratory activities from Ing. Petr Bauer, is acknowledged.

Conflicts of Interest: The authors declare no conflict of interest.

References

1. Kadeřávek, P.; Pernička, J. Torsion oscillations of powered wheelsets. *Railvolution* **2013**, *7*, 34–37.
2. Anand, P. Coning of Wheels in Railways: Theory and Purpose. Available online: https://civildigital.com/coning-wheels-railways-theory-purpose/ (accessed on 22 June 2023).
3. Markovic, P.; Kostic, D.; Bojovic, N. One Method for Detection of Torsional Oscillations of Driving Axles of Electrical Locomotives. In Proceedings of the PRORAIL 2015—XXII. Medzinárodná Konferencia Súčasné Problémy v Koľajových Vozidlách, Žilina, Slovakia, 16–18 September 2015; Žilinská Universita, Strojnická Fakulta: Žilina, Slovakia, 2015; Volume 2, pp. 27–37.
4. Trimpe, F.; Lück, S.; Naumann, R.; Salander, C. Simulation of Torsional Vibration of Driven Railway Wheelsets Respecting the Drive Control Response on the Vibration Excitation in the Wheel-Rail Contact Point. *Vibration* **2020**, *4*, 30–48. [CrossRef]
5. Kondo, K. Anti-slip Control Technologies For the Railway Vehicle Traction. In Proceedings of the 2012 IEEE Vehicle Power and Propulsion Conference, Seoul, Republic of Korea, 9–12 October 2012; pp. 1306–1311. [CrossRef]
6. Sebesan, I.; Mitu, A.; Sireteanu, T. On the stick-slip phenomena in traction railway vehicles. *Proc. Rom. Acad. Ser. A* **2015**, *16*, 209–2016.
7. Wu, G.; Shen, L.; Yao, Y.; Song, W.; Huang, J. Determination of the dynamic characteristics of locomotive drive systems under re-adhesion conditions using wheel slip controller. *J. Zhejiang Univ. A* **2023**, *24*, 722–734. [CrossRef]
8. Reitmeier, D.; Mertens, A. Active Reduction of Gear Mesh Vibrations by Drive Torque Control. In Proceedings of the 2023 IEEE International Conference on Mechatronics (ICM), Loughborough, UK, 15–17 March 2023. [CrossRef]
9. Fleischer, M. Modal State Control in the Frequency Domain for Active Damping of Mechanical Vibrations in Traction Drive-Trains. In Proceedings of the 8th IEEE International Workshop on Advanced Motion Control, 2004. AMC '04, Kawasaki, Japan, 28 March 2004.
10. Abouzeid, A.; Trimpe, F.; Lück, S.; Traupe, M.; Guerrero, J.; Briz, F. Co-Simulation-Based Verification of Torsional Vibration Protection of Electric-Driven Railway Vehicle Wheelsets. *Vibration* **2020**, *5*, 613–627. [CrossRef]
11. Sadr, S.; Khaburi, D.; Rodriguez, J. Predictive Slip Control for Electrical Trains. *IEEE Trans. Ind. Electron.* **2016**, *63*, 3446–3457. [CrossRef]
12. Hajano, F.; Abro, F.; Memon, T.; Kalwar, I.; Burhan. Fuzzy Logic based Anti-Slip Control of Commuter Train with FPGA Implementation. *Int. J. Adv. Comput. Sci. Appl.* **2020**, *11*, 293–300. [CrossRef]

13. Breuer, W.; Yu, M. Energie-Methode: Vorhersage des maximalen dynamischen Torsionsmomentes. In Proceedings of the 16. Internationale Schienenfahrzeugtagung Dresden, Dresden, Germany, 12–14 September 2018.
14. Liu, J.; Zhao, H.; Zhai, W. Mechanism of self-excited torsional vibration of locomotive driving system. *Front. Mech. Eng. China* **2010**, *5*, 465–469. [CrossRef]
15. Qi, Y.; Dai, H. Influence of motor harmonic torque on wheel wear in high-speed trains. *J. Rain Rapid Transit* **2020**, *234*, 32–42. [CrossRef]
16. Spangenberg, U.; Fröhling, R. Solving locomotive wheel polygonization by reducing variable frequency drive interharmonics. *Proc. Inst. Mech. Eng. Part F J. Rail Rapid Transit* **2021**, *235*, 73–82. [CrossRef]
17. Winterling, M. Oscillations in Rail Vehicle Traction Drives, Analysis of Electromechanics. Ph.D. Thesis, Delft University, Delft, The Netherlands, 1997; p. 193.
18. Szolc, T.; Konowrocki, R.; Michajlow, M.; Pregowska, A. An investigation of the dynamic electromechanical coupling effects in machine drive systems driven by asynchronous motors. *Mech. Syst. Signal Process.* **2014**, *49*, 118–134. [CrossRef]
19. Schneider, R. Torsionsschwingungen von Radsatzwellen—Systemanalyse Teil 1: System-und Modellbeschreibung. *ZEVrail* **2017**, *141*, 452–461.
20. Schneider, R. Torsionsschwingungen von Radsatzwellen—Systemanalyse Teil 2: Physikalische Untersuchungen und Sicherheitsbetrachtung. *ZEVrail* **2018**, *142*, 27–39.
21. Takahashi, A.; Sugimoto, S.; Tamiya, S.; Fujii, K.; Endo, M. High-Efficiency Technology for Railway Vehicle Traction Motors. In Proceedings of the 23rd International Conference on Electrical Machines and Systems (ICEMS), Hamamatsu, Japan, 24–27 November 2020; pp. 2114–2117. [CrossRef]
22. Lu, B.; Liu, Z.; Wang, X.; Song, Y.; Zhang, Q.; Wu, S.; Tao, G. Influence of Electric Traction Drive System Harmonics and Interharmonics on the Vibration of Key Locomotive Components. *IEEE Trans. Veh. Technol.* **2023**, *72*, 12830–12844. [CrossRef]
23. Švestka, D. Atlas Locomotive. Available online: http://www.atlaslokomotiv.net/loko-380.html (accessed on 7 March 2024).
24. Dybala, V.; Šulc, B. Simulation of Influence of Transverse Wheel-Set Movement on Torsion Oscillations. In *Sborník Vědeckých Prací Vysoké Školy Báňské—Technické Univerzity Ostrava, Řada Strojní*; VŠB - Technická univerzita Ostrava: Ostrava, Czech Republic, 2019.
25. Kolář, J. Design of a Wheelset Drive. *Trans. Electr. Eng.* **2015**, *4*, 11–19.
26. Dybala, V. Simulace zvlnění elektromagnetického momentu asynchronního trakčního motoru lokomotivy napájeného ze střídače. In *Sborník Konference STČ na Strojní Fakultě ČVUT v Praze*; ČVUT v Praze, Fakulta Strojní: Praha, Czech Republic, 2019.
27. Polach, O. Creep forces in simulations of traction vehicles running on adhesion limit. In Proceedings of the 6th International Conference on Contact Mechanics and Wear of Rail/Wheel Systems, Göthenburg, Sweden, 10–13 June 2003.
28. Javůrek, J. *Regulace Moderních Elektrických Pohonů*; Grada Publishing, a.s.: Praha, Czech Republic, 2003.
29. Novák, J.; Chyský, J. Vyšetřování ztrát sinusových filtrů pro frekvenčně řízení pohony. In Proceedings of the XXXIII. Celostátní konference o Elektrických Pohonech, Plzeň, Czech Republic, 11 – 13 June 2013.
30. Dybala, V. The electromagnetically excited resonance of the pinion in fully-suspended drive of a locomotive and its sensitivity on the rotor shaft stiffness. In Proceedings of the STČ 2021—Konference Studentské Tvůrčí Činnosti, Praha, Czech Republic, 4 May 2021.
31. Fridrichovský, T. Vztah mechanických a elektrických veličin v pohonech moderních kolejových vozidel. In *Současné Problémy V Kolejových Vozidlech 2021: XXV. Konference S Mezinárodní Účastí, Sborník Příspěvků*; Univerzita Pardubice, Dopravní Fakulta Jana Pernera: Česká Třebová, Czech Republic, 2021; pp. 55–62.

vehicles

MDPI

Article

Development of a Design Procedure Combining Topological Optimization and a Multibody Environment: Application to a Tram Motor Bogie Frame

Alessio Cascino *, Enrico Meli and Andrea Rindi

Department of Industrial Engineering, University of Florence, 50139 Florence, Italy; enrico.meli@unifi.it (E.M.); andrea.rindi@unifi.it (A.R.)
* Correspondence: alessio.cascino@unifi.it

Abstract: Nowadays, it is essential to find increasingly rapid and efficient design strategies. This approach becomes crucial in the railway industry, where components must be verified according to multiple reference standards, both structurally and dynamically. In this context, the present research activity aims to develop a fast and effective desin procedure based on European reference standards. The goal was to develop the geometry of a motor bogie frame for a tram vehicle, integrating three fundamental tools for development: finite element simulations and topological structural optimization, a Write Computer Aided Design (CAD) environment, and a multibody environment. Their integration could enhance design accuracy, streamline the traditional design workflow, and support innovation. The optimization process involved the introduction of complex technological constraints, directing the geometry toward production by casting. A tool was developed to automate running dynamics simulations and the output of results for immediate verification of the entire vehicle performance. Finally, the new geometry was tested both structurally and dynamically. The mass was reduced by approximately 7% while ensuring satisfactory mechanical performance. The maximum value of stress was reduced by about 16%. The dynamic performance showed negligible variation, confirming the encouraging outcomes to make this procedure increasingly effective and reliable.

Keywords: railway vehicles; structural optimization; topological optimization; multibody dynamics; manufacturing constraints; railway bogie

1. Introduction

Nowadays, railway vehicles stand out as one of the optimal solutions for sustainable mobility, aiming to reduce air pollution generated by various types of land and non-land vehicles. The growth of the railway sector necessitates addressing new challenges in both vehicle design and management. A critical aspect of this involves the integration of running dynamics analysis with structural optimization to enhance performance and safety. Running dynamics analysis is crucial for understanding the operational behavior of railway vehicles, leading to improved stability and passenger comfort. When combined with structural optimization, this approach allows for the design of components that are not only lightweight and durable but also capable of maintaining optimal performance under varying operational conditions. This synergy can lead to significant advancements in the efficiency and reliability of railway systems. The focus of this research activity was the innovation of a motor bogie mounted on a tram vehicle. Redesign of this component was achieved through a combination of key tools used in the railway sector for the development of such components: the creation of a high-level Finite Element (FE) model, the addition of a complex structural optimization calculation, the reconstruction of a CAD model, and the development of a multibody model. The ability to effectively and automatically integrate these environments while adhering to all relevant European design standards represents the novelty of this activity, along with the innovative design and validation of the bogie. Structural optimization processes have yet to establish themselves in the railway field, especially as an optimization method. The present research activity has confirmed the potential of this tool in which technological constraints were also included to ensure that the optimized geometry could be produced using a sand-casting process. The results were excellent, as was the impact of a sensitivity analysis, which significantly reduced computation times. Finally, the automation of multibody analyses to study the running dynamics of the entire vehicle, including optimized geometries, further increase the efficiency of the design process. Having a replicable bogie frame model would lead to a significant reduction in production times, both during initial construction and when repairs or component replacements are needed due to damage. In order to reach these objectives, topological structural optimization processes are essential tools that support the design and development phases of the system in question. The literature highlights that topological structural optimization approaches have yet to gain widespread acceptance in the railway field, as well as a design procedure capable of efficiently integrating an FE model, a CAD model, and a multibody model. This aspect is even more accentuated for tram vehicles, which are the focus of the present work. In general, topological and structural optimization processes are effective techniques for designing various types of components. A suspension system adopted in automotive applications has been studied to emphasize a weight reduction [1]. An automotive car body was optimized through material performance indices [2], and a high-performance chassis design of an automotive vehicle body has been innovated using topological optimization [3]. Comparable methods have been applied to enhance the design process for turbomachinery components [4], particularly when integrated with additive manufacturing techniques [5–8]. Moving to railway applications, the modal behavior of a car body structure was included in a size optimization approach testing an innovative design approach [9]. In addition, composite structures were integrated and optimized, testing their structural and dynamic behavior [10]. Another similar activity regarded an anchor bracket, developed using topological optimization applied on composite materials [11] and a composite wagon [12]. A new generation of freight railway wagons was developed and proposed in [13]. Numerous researchers focused on optimizing car bodies with an emphasis on lightweight design processes [14], using a material selection method [15] or a multidisciplinary approach [16]. A recent optimization application based on a sensitivity analysis for a railway vehicle body is presented in [17]. With reference to manufacturing constraints, the bolster beam and the bogie frame of a freight wagon were optimized to reduce the mass of the system [18]. A topological optimization was conducted using Solid Isotropic Material with Penalization (SIMP) methodology to maximize the stiffness of the structure. Two casted parts were compared, highlighting the impact of manufacturing constraints in the optimization process [19]. Additionally, the existing literature emphasizes the critical importance of accurately predicting the lifespan of railway bogie frames. An FE model and multibody dynamics model were simulated to study the behavior of bogie frames under real operational service loads [20]. A similar approach was adopted to test the crashworthiness performance of a passenger rail vehicle locomotive [21]. The dynamic response acting on a high-speed train was analyzed using a rigid-flexible coupled vehicle dynamic model [22]. The bogie frame of a diesel multiple unit (DMU) was studied using stress and acceleration tests to clarify the reasons for fatigue cracking [23]. With reference to railway dynamics in general, a dynamic model for an asymmetrical vehicle/track system was proposed in [24], while the performance of a pantograph–catenary system was analyzed in [25]. However, all the proposed techniques focused on evaluating vehicle performance, neglecting the aspects of component redesign and innovation. It confirms the importance of the novelty proposed in this research activity, where a complex component like a tramway bogie frame was innovated combining all the main techniques for component development. The process resulted in an enhancement of the traditional procedure, providing a solid foundation for direct integration between structural optimization and the multibody environment, with new applications of a flexible multibody. The structure of this paper is summarized as follows: Section 1 includes an introduction to the research work and a literature review. Section 2 presents the methodology, and the CAD and FE models of the bogie, and the multibody model of the whole vehicle. Section 3 outlines the results obtained from the proposed procedure, focusing on structural, vibrational, and running dynamics aspects. Finally, Section 4 presents the conclusions and potential future developments.

2. Methodology and Benchmark Description

The following section provides a comprehensive overview of the methodology utilized in this study. Along with a brief description of the tram bogie frame, including an FE model, the optimization process and a multibody model are also presented.

2.1. Model Description: The Metro Bogie Frame

The tram bogie frame serves as a critical structural component, engineered to support and integrate various elements within the train undercarriage. In detail, the tram vehicle tested during the activity (P-V category according to EN 12663-1:2010+A2:2023 [26]) had five car bodies, two motor bogies, and a trained one. The motor bogie frame was studied and innovated. This robust frame is designed to deliver the strength, durability, and resilience necessary to withstand the rigorous operational demands of a tram system. Precision welding techniques are employed in the fabrication of the frame, ensuring seamless and uniform connections between steel components. Despite precision in manufacturing, welded joints remain a significant concern, particularly regarding mechanical fatigue acting in these complex systems. The frame comprises two robust longitudinal beams, two crossmembers, and other geometric configurations, strategically designed to evenly distribute loads across the structure. This design not only enhances the stability of the bogie but also contributes to the overall safety and efficiency of the tram vehicle system. The original bogie frame was constructed using structural steel and assembled via welding as described. In the middle part of the longitudinal beam are positioned secondary suspension interfaces. At the ends, in opposite positions, supports for the motor and braking system are located. The connection interfaces of the two traction rods are positioned on the two crossbeams. Figure 1 provides a comprehensive view of the complete frame.

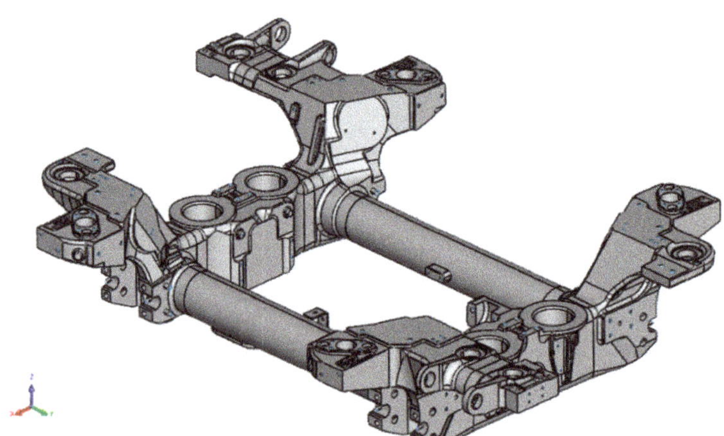

Figure 1. View of the original tram bogie frame.

2.2. Methodology

The methodology adopted by the authors for the innovation of a tram motor bogie frame is outlined in detail and summed up in Figure 2. Several steps were needed to achieve an innovated configuration that complied with structural and running dynamics requirements. Different environments were involved. FE calculation was fundamental both for static-fatigue structural assessment and for the optimization process. A CAD environment was useful for transitioning from a design concept to a tangible design. Finally, a multibody ambient allowed virtual testing of the complex dynamics of the vehicle in different running conditions. The possibility of combining these tools represents a novelty in the railway field, allowing fast management and checking, which are main design aspects of a component as complex as a railway bogie frame. In addition, the present research work represents the development of an interesting design methodology approach proposed by authors in [27]. It involved the implementation of a robust sensitivity analysis on the loading conditions to which railway bogies are subjected to expedite the complex topological optimization process, which also included manufacturing constraints. Returning to the present research activity, the first step of the procedure involves assessing the original performance of the component to be innovated, both from a structural perspective and in terms of dynamic behavior. Once these are known, the process moves on to the first and only step of topological optimization. This step enables the development of an innovative frame design, based on load conditions proposed by the European standard EN 13749:2021+A1:2023 [28]. This standard includes approximately 20 load conditions, including both static and fatigue-related conditions, summarized in a reference table found in [27]. Information regarding optimization process settings is also included in this reference. Furthermore, the optimization process incorporates manufacturing constraints with the goal of producing a design suitable for casting, thereby eliminating the most critical element of this component: welds. Proceeding to the central block, which includes an automated loop, the first step involves reconstructing the geometry resulting from optimization using commercial CAD software. At this point, a quick modal analysis of the bogie frame in free-free conditions could also be performed to observe its natural vibration frequencies. This is not a mandatory step. The critical frequencies of the bogie from a dynamic behavior perspective are quite low; for example, hunting typically occurs at around 6 Hz. In contrast, bogie frames usually exhibit frequencies well above 40 Hz. However, during the final phase of analysis for the completed frame, a comparison of frequencies should always be conducted to fully conclude the analysis of the innovated component. Therefore, once the initial version of the geometry is completed, through a tool developed by the authors using Python (version 3.12), the data related to the new design automatically update the input file of the multibody software used, specifically Simpack 2021. The model consists solely of rigid bodies, including the bogie frame. Therefore, it is sufficient to update the component data related to moments of inertia, mass, and center of gravity position. Once the new Simpack model is set up, the simulation will automatically start. The final step of the automated cycle is the production of results, where key dynamic performance parameters, such as the vehicle critical speed, axle lateral displacement, derailment coefficient, and normal and tangential pressures, are compared. If the parameters are satisfactory, the dynamic behavior is also confirmed. This allows for moving on to the final testing phase, where static and fatigue mechanical performance are re-evaluated. Is it possible for the new design to exhibit some critical areas based on the material allowable values? Yes, it is possible. However, this will not necessitate a new dynamic test of the entire vehicle for two main reasons. First, the innovative design is the result of an optimization process in which the load conditions have already been considered for strength purposes. Second, as a consequence of the first point, stress concentrations may occur in very localized portions of the material. Therefore, corrective design actions will also be localized, without affecting the dynamic behavior of the entire vehicle system.

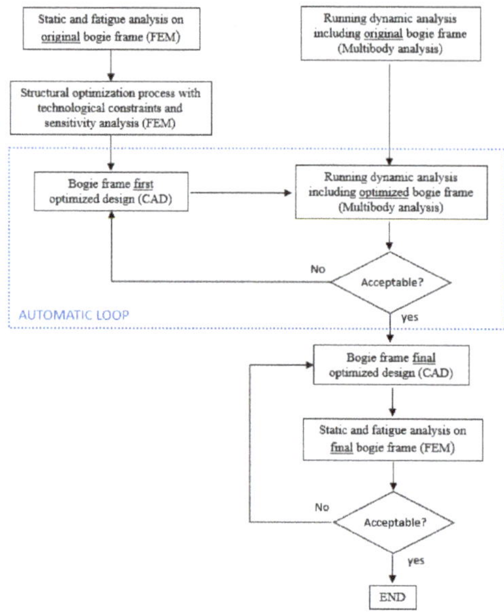

Figure 2. Main steps of the methodology.

2.3. FE Model of the Motor Bogie Frame

Figure 3 illustrates the FE model of the original motor bogie frame. It was a high-level representation of the system, where every part was modeled with different types of elements (0D, 1D, and 3D), according to component requirements. Due to the robust geometry of the frame, characterized by thick sections and three-dimensional development geometries, it was not feasible to represent it using the mathematical formulation of 2D shell elements (QUAD type). All the main components, including longitudinal beams, cross members, and various supports, were modeled with a three-dimensional mesh of second-order TETRA elements (TETRA10 type), adapting the element dimension to the various parts of the bogie. The choice of different types of elements depended on the characteristics of the modeled components. Figure 4 shows a detailed view of the mesh. All the supports were connected to the main frame, using contacts formulation, specifically using freeze/bonded contact. This type of contact configuration imposes the linearity of the model, and the correct redistribution of all the acting stresses (both forces and moments). It enforces zero relative motion on the contact interfaces, and the gap between them remains fixed at the original value fixed by the user. The sliding distance is forced to be zero. This choice allowed accuracy of the results and simplicity of the modeling. Moreover, the model is easier to manage as the various components can be modified independently. It is sufficient to maintain the interface between the components to ensure the connection remains active, expediting potential changes and numerical tests. Depending on the type of loads and the zone of application, forces, pressures, and RBE3 elements were used to load the bogie frame. Using RBE3 elements allowed exclusion of the addition of stiffness to the model, distributing force and moments among the selected nodes. The wheelsets were represented with one-dimensional elements. In detail, a series of connected beams with the proper section dimensions and shape were used. This modeling choice gave the right mass and inertial characteristics to the model, without losing the flexural behavior of the component. The primary suspension (light blue color) mounted between the wheelset and the bogie frame was also modeled using one-dimensional elements, following the same concept described above. Accurately representing this system was critically important as it was vital to precisely replicate this connection point. It was necessary to avoid introducing overly rigid constraints that could create excessively high stresses, leading to unrealistic results. Global constraint conditions applied to the bogie frame; all load cases were represented by an isostatic configuration. The reference points were the four wheels, positioned at the end of the axle.

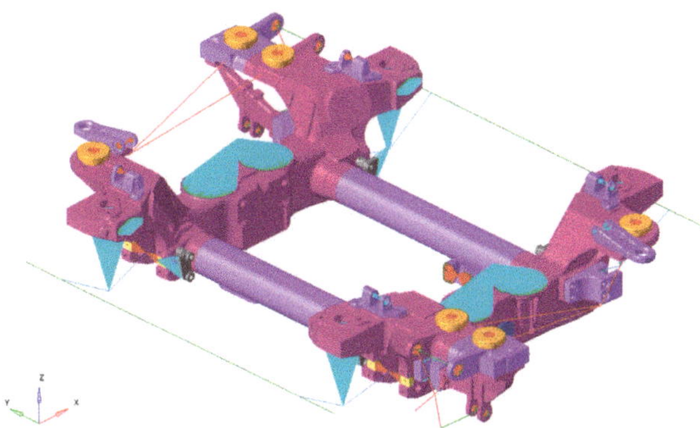

Figure 3. FE model of the original motor bogie frame.

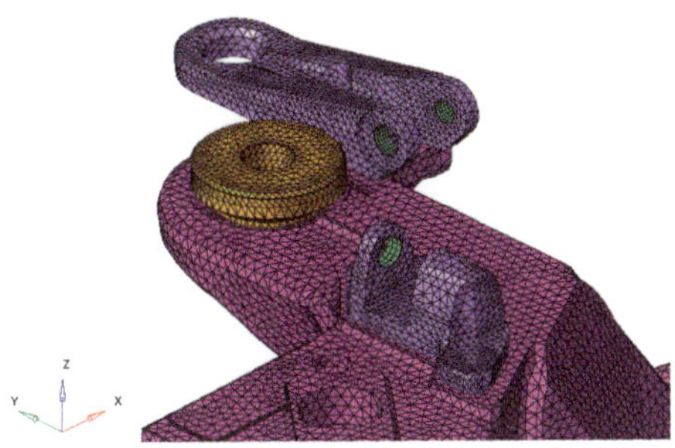

Figure 4. Mesh detail of the FE model.

2.4. Topological Optimization Model and Settings

The topological optimization process applied to the tram motor bogie frame during the present research activity had the objective to propose an innovative design for the frame, which would be suitable for a sand-casting production process. All information about settings is detailed in [27] as explained in Section 2.2. The solver was based on the Gradient based Optimization Method [29]. All simulations were conducted using the same computer that had the following characteristics: Intel(R) Xeon(R) CPU E5-2643 v4 @ 3.40 GHz and RAM 32 GB. The goal of the optimization process was to minimize the weighted compliance evaluated across all load cases. The compliance represents the sum of strain energy of the mechanical model. The higher the value, the more the component is deformed under a load. This essentially establishes an inverse measure of stiffness, enabling the formulation of a more efficient minimization problem. Before proceeding with the optimization process, the model was adapted to the methodology (Figure 5). The volume of modifiable material was expanded, giving the solver greater flexibility (green color), while the interface and connection areas with other vehicle components, the connection areas with various supports, and the load application zones were included in the non-modifiable material volume (red color). The optimization constraint was imposed on the mass fraction, which is the ratio between the final and initial mass of the modifiable material (design space). It was set below 40%, which was the reference value adopted for topological optimization.

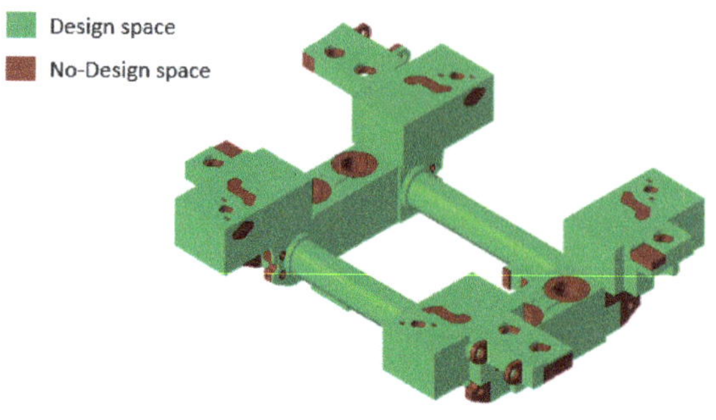

Figure 5. Topological optimization model of the motor bogie: design and non-design space.

2.5. Multibody Model

The multibody model used represents a five-coach tram vehicle, with three coaches supported by bogies and two suspended. The end bogies are motorized, while the central one is a trailing bogie. All components of the vehicle have been modeled as rigid bodies. In addition to the coaches, which are clearly visible in Figure 6a, all the fundamental elements of each bogie have been modeled as shown in Figure 6b. Furthermore, the characteristics of the bogie frame are also included as fundamental parameters: mass, centroidal moments of inertia, and the position of the center of mass. The two wheelsets present on each bogie are also clearly visible. Finally, the rails were of the "grooved" type, which is consistent with the nature of an urban vehicle.

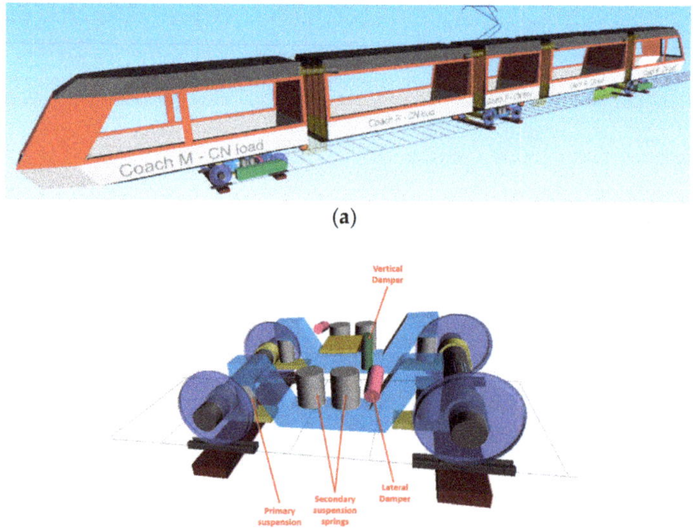

(a)

(b)

Figure 6. Multibody model: (**a**) global view of the vehicle, (**b**) detailed view of the bogie.

3. Results and Discussions

3.1. Topological Optimizations Result and Innovative Bogie Frame Design

Once the optimization process was completed, an initial material density distribution was obtained, as shown in Figure 7. The red zones represent areas where the material is necessary and cannot be removed to ensure the required performance of the frame in accordance with optimization process settings. Moving towards the lower end of the scale, represented in blue is material that can be removed. As shown in the reference figure, the solver generated large cavities both in the longitudinal beams and crossbeams. This significant result positively directs the component toward the sand-casting production process thanks to the technological constraints introduced in the optimization calculation. Figure 8 shows the trend of the objective function during the optimization process, highlighting the 7% mass savings achieved.

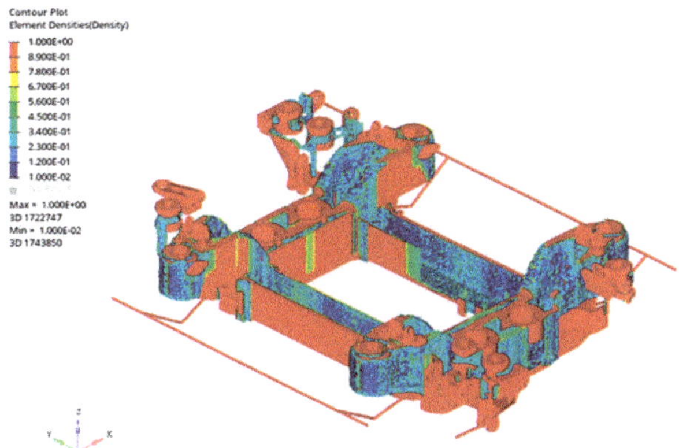

Figure 7. Topological optimization result: element density distribution.

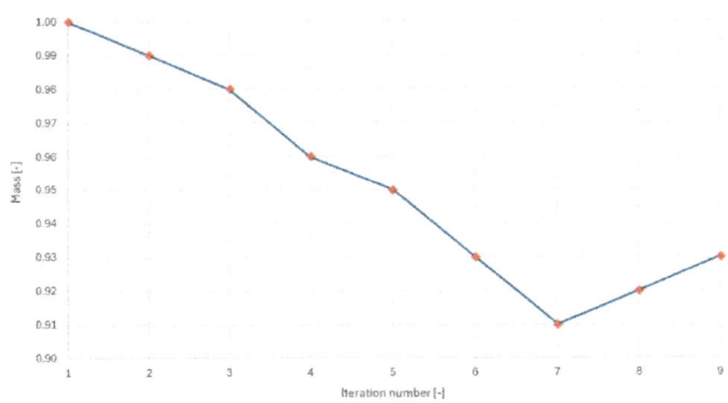

Figure 8. Objective function of the optimization process.

The next step involved reconstructing the geometry of the frame as shown in Figure 9. This new design is highly innovative compared to more conventional geometries, such as the original one. The bodies of the longitudinal beams are hollow as are those of the crossbeams. The crossbeams also feature additional internal reinforcements that ensure the strength and stiffness of the entire frame. Moreover, their presence enhances the casting process by creating a much better flowability condition for the molten material. The two crossbeams have their openings facing downward. This design choice proved to be crucial as it prevents the accumulation of debris during vehicle operation, thereby reducing potential sources of damage. The interface area of the secondary suspensions includes an additional cavity. The supports have retained their geometry at the interface zones, while being globally streamlined and adapted to the new production process, with larger fillet radii and smoother shapes. The large central holes in the longitudinal beams allow for a significant reduction in mass but require the addition of cores to the mold used for production.

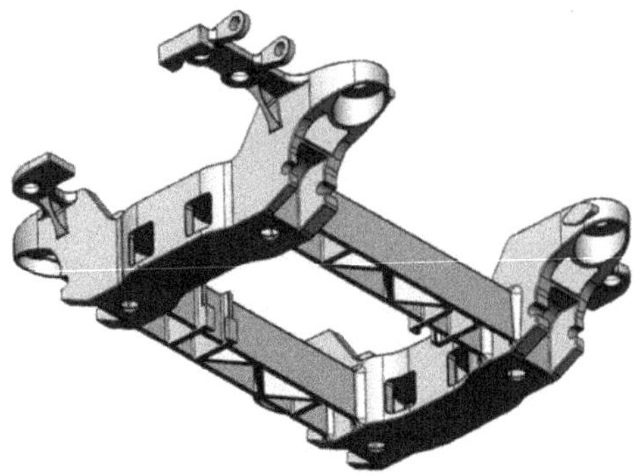

Figure 9. Innovative motor bogie frame (CAD view).

3.2. Bogie Frame: Innovative Design and Verification

The mass and inertial properties of the new bogie frame were then automatically incorporated into the vehicle Simpack model. The objective of this step was to observe minimal variations in the dynamic behavior of the entire vehicle, now supported by two newly designed motor bogies with different mechanical and vibrational properties. A modal analysis of the frame under free-free conditions was conducted, confirming the robustness of its vibrational performance. However, the results will be presented later as the test was repeated on the final geometry. Two test conditions were considered: straight track and 100 m curve track. The comparison between the original model and the current one was conducted based on several reference parameters, including the vehicle critical speed, lateral displacement of the wheelsets, Y/Q coefficient (also known as the derailment coefficient), and normal and tangential forces. The critical speed of a railway vehicle is the speed at which the dynamic behavior of the vehicle becomes unstable, potentially leading to unsafe conditions. At this speed, the interaction between the vehicle and the track, particularly the forces transmitted through the wheel–rail interface, can result in excessive vibrations, lateral oscillations, or even derailment. The phenomenon is closely related to the vehicle suspension system, track geometry, and the natural frequencies of the vehicle components. Critical speed is important because it defines the operational limits for the safe and efficient functioning of railway systems. Exceeding this speed can compromise passenger comfort, increase wear and tear on both the vehicle and track, and, in extreme cases, pose serious safety risks. In this case, the critical speed of the vehicle has remained unchanged and is approximately 90 km/h, confirming a positive result. The derailment coefficient (Y/Q) is a crucial parameter in railway dynamics that represents the ratio between the lateral force (Y) and the vertical load (Q) exerted on a railway wheel. Specifically, the lateral force (Y) is the force acting horizontally at the wheel–rail interface, while the vertical load (Q) is the downward force due to the weight of the vehicle. The Y/Q ratio is used to assess the risk of wheel derailment, which occurs when the lateral forces exceed a critical threshold relative to the vertical load. By comparing the two curves obtained for the curved track, no significant differences are observed, confirming only a minimal variation in vehicle dynamic behavior. Additionally, the maximum recorded value of the derailment coefficient remains below the permissible limit of 1.2. This indicates that the vehicle operates safely within acceptable thresholds, with no increased risk of derailment even when navigating curved sections of the track. The derailment coefficient is important because it directly influences the safety of railway operations. A high Y/Q ratio indicates that the lateral forces are approaching a level that could lift the wheel off the rail, potentially leading to derailment. Figure 10 shows the result for the Y/Q coefficient, obtained with a 20 Hz low-pass filter.

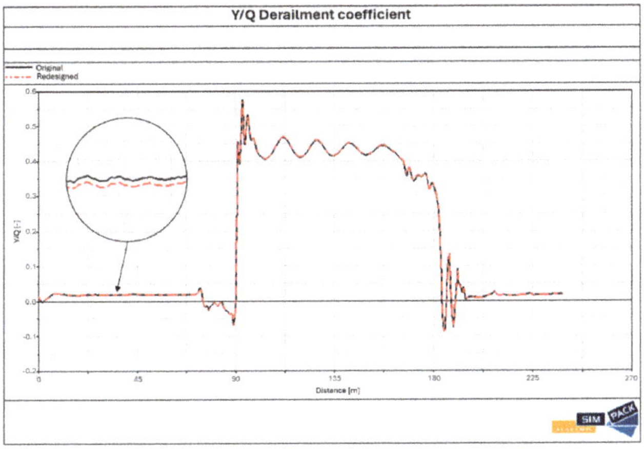

Figure 10. Comparison between original and redesign model: Y/Q derailment coefficient.

For all other parameters, similar results were obtained, with negligible differences compared to the original. Therefore, the representations of these parameters have been omitted. Given the results, the proposed chassis has been deemed suitable.

3.3. Structural Performance Assessment

The bogie frame was tested again for fatigue under load conditions specified by the reference standard. Specifically, seven static conditions and nine fatigue conditions were analyzed. The FE model of the bogie frame was tested by verifying the performance in terms of stress at each node of the structure grid. Starting from the local stress tensor, it was possible to determine the resultant of the principal stress components. The values obtained were compared with the reference allowable limits. Before proceeding with the calculation, the model was subjected to a sensitivity analysis that took into account the effects of mesh size on the maximum stress value. The results are shown in Figure 11, where it can be observed that variation in the normalized maximum stress value is less than 3% after the last two iterations, confirming the stability of the FE model.

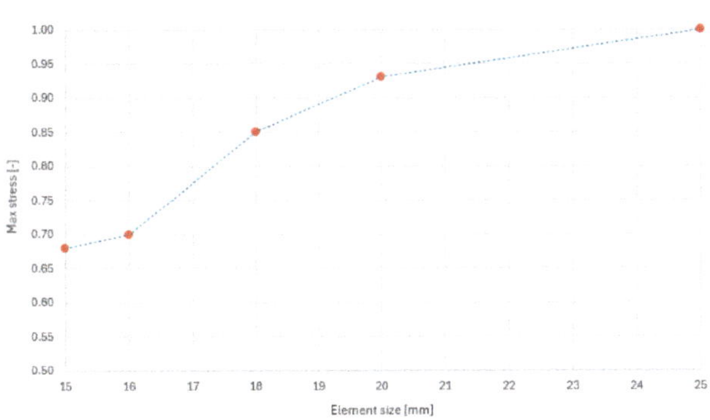

Figure 11. Sensitivity analysis on mesh element size.

Table 1 summarizes the results obtained from fatigue calculation, exploiting the utilization coefficient of the material, defined as the ratio between the stress calculated value and permissible value. The results were positive, with only one critical area characterized by a utilization coefficient larger than one, identified at the base of a support on the extremity of the longitudinal beam (as indicated by the square in Figure 12). Stress values have been normalized to the maximum observed in the original bogie configuration. The stress permissible value has been expressed as a range.

Table 1. Fatigue analysis results.

Δσ (Principal Stress) [-]		**Δσ (Permissible Stress Range) [MPa]**	**Utilization Factor (U) [-]**	
Original	**Innovated**	**-**	**Original**	**Innovated**
1	0.84	130–150	1.43	1.20
0.81	0.68	130–150	1.16	0.97

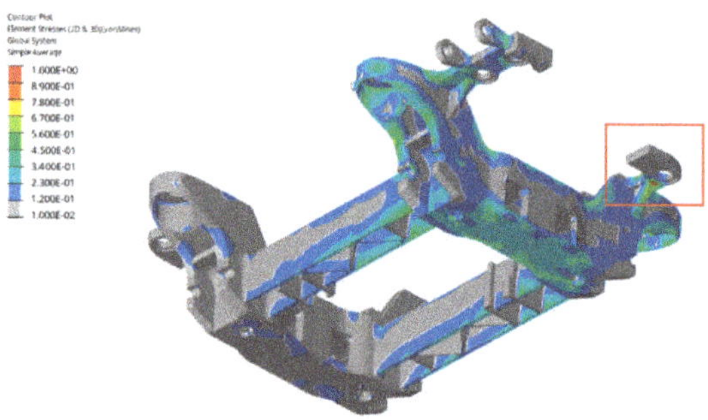

Figure 12. Stress distribution visualization of the innovated motor bogie frame.

The maximum allowed value in linear analysis is equal to one. Local exceedances of stress are allowed, taking into account the plastic behavior that characterizes metallic materials. This research activity did not require it, but a non-linear calculation could be exploited to investigate the plastic behavior of the system.

From the perspective of the dynamic behavior of the innovative frame, a comparison with the original frame is presented. This allows us to observe that the first mode of vibration has remained essentially unchanged. As shown in Figure 13, the dark blue areas are more extensive in the new proposed solution, confirming an overall increase in stiffness in the central area of the bogie. The first mode is a flexural mode that locally involves one of the end supports. This analysis was crucial, both to confirm that the first natural frequency of the system exceeds 50 Hz and to identify the area of maximum displacement, which coincides with the peak stress region discussed earlier. Finally, Figure 14 presents a comparison of the first six natural frequencies of the two frames. The match in the first frequency is excellent and nearly identical to that of the original frame, confirming the positive outcome of the analysis.

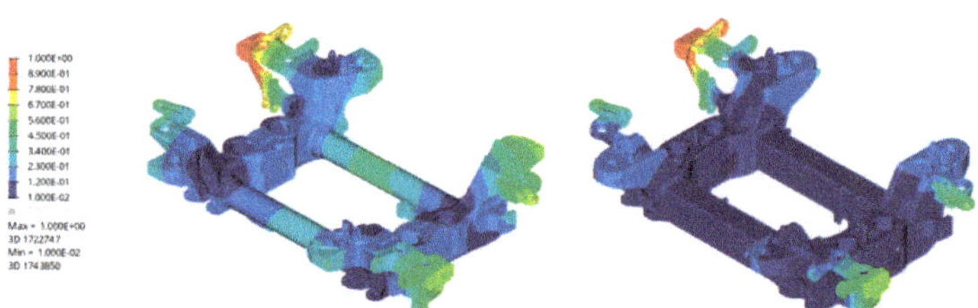

Figure 13. Mode shapes comparison for mode 1: original frame on the **left** and innovated frame on the **right**.

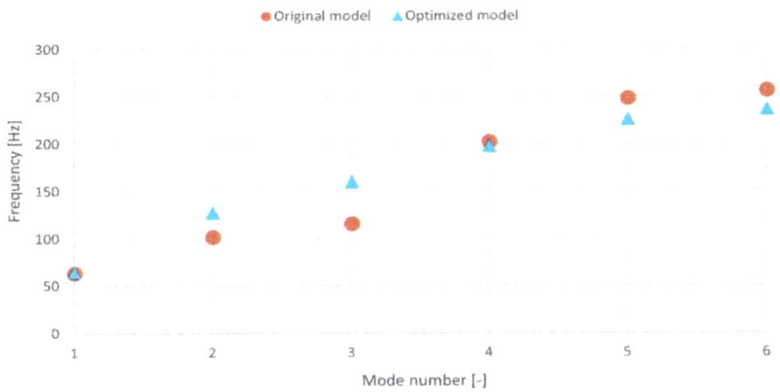

Figure 14. Modal frequencies comparison.

4. Conclusions and Future Developments

This project involved the combination of several essential tools for the design of a railway bogie. Starting with a high-level FEM model of a tram motor bogie, it was subjected to a structural optimization process. This approach, whose load conditions were based on the relevant European standard for bogie design, enabled the development of an innovative frame design. Additionally, due to the proposed technological constraints, the frame proved to be highly innovative and could be produced through a sand-casting process. In parallel, a multibody model of the vehicle was created to test its running dynamics. The goal of this project was to efficiently integrate these methodologies, effectively and quickly, to develop an innovative bogie frame design that considers both structural and vibrational performance as well as the dynamics of the entire vehicle. The innovative design demonstrated good performance. The mass was reduced by approximately 7% despite the transition to a new manufacturing technique. From both a static and fatigue perspective, it presented only one critical area, which was easily resolved. The first vibration mode maintained both its frequency and modal shape, confirming a first bending mode that locally involved an end support. Finally, the change in mass properties of the new frame did not cause differences in terms of the vehicle critical speed (running stability) or derailment coefficient (Y/Q). Other analyzed parameters also confirmed this positive result. This first phase of the research activity allowed for the development of a satisfactory approach to generating a new design. As a future development, this activity aims to create an automatic loop capable of integrating the FE model of the bogie and the optimization environment directly with the multibody environment, incorporating flexible behavior. In addition, the optimization process itself will need to be improved, with the introduction of new constraint conditions related to mechanical fatigue.

Author Contributions: Conceptualization, methodology, software, simulation, and validation, A.C.; review, supervision, and project administration, E.M. and A.R. All authors have read and agreed to the published version of the manuscript.

Funding: This study was supported by MOST Sustainable Mobility National Research Center and received funding from the European Union Next GenerationEU (PIANO NAZIONALE DI RIPRESA E RESILIENZA (PNRR) MISSIONE 4 COMPONENTE 2, INVESTIMENTO 1.4 D.D. 1033 17/06/2022, CN00000023).

Data Availability Statement: Data is unavailable due to privacy.

Acknowledgments: All the figures have been provided by Hitachi Rail S.p.A (Piossasco, Italy).

Conflicts of Interest: The authors declare no conflicts of interest.

References

1. Kong, Y.; Abdullah, S.; Omar, M.; Haris, S. Topological and Topographical Optimization of Automotive Spring Lower Seat. *Lat. Am. J. Solids Struct.* **2016**, *13*, 1388–1405. [CrossRef]
2. Cui, X.; Zhang, H.; Wang, S.; Zhang, L.; Ko, J. Design of lightweight multi-material automotive bodies using new material performance indices of thin-walled beams for the material selection with crashworthiness consideration. *Mater. Des.* **2011**, *32*, 815–821. [CrossRef]
3. Cavazzuti, M.; Baldini, A.; Bertocchi, E.; Costi, D.; Torricelli, E.; Moruzzi, P. High performance automotive chassis design: A topology optimization based approach. *Struct. Multidiscip. Optim.* **2011**, *44*, 45–56. [CrossRef]
4. Bandini, A.; Cascino, A.; Meli, E.; Pinelli, L.; Marconcini, M. Improving Aeromechanical Performance of Compressor Rotor Blisk with Topology Optimization. *Energies* **2024**, *17*, 1883. [CrossRef]
5. Pietropaoli, M.; Ahlfeld, R.; Montomoli, F.; Ciani, A.; D'Ercole, M. Design for Additive Manufacturing: Internal Channel Optimization. In Proceedings of the ASME Turbo Expo 2016: Turbomachinery Technical Conference and Exposition. Volume 5B: Heat Transfer, Seoul, Republic of Korea, 13–17 June 2016; ASME: New York, NY, USA, 2016. V05BT11A013.
6. Gardan, N.; Schneider, A. Topological optimization of internal patterns and support in additive manufacturing. *J. Manuf. Syst.* **2015**, *37*, 417–425. [CrossRef]
7. Gebisa, A.W.; Lemu, H.G. A case study on topology optimized design for additive manufacturing. In Proceedings of the IOP Conference Series: Materials Science and Engineering, Busan, Republic of Korea, 25–27 August 2017; Volume 276, p. 012026.
8. Ranjan, R.; Samant, R.; Anand, S. Integration of design for manufacturing methods with topology optimization in additive manufacturing. *J. Manuf. Sci. Eng.* **2017**, *139*, 061007. [CrossRef]
9. Cascino, A.; Meli, E.; Rindi, A. Dynamic size optimization approach to support railway carbody lightweight design process. *Proc. Inst. Mech. Eng. Part F J. Rail Rapid Transit* **2023**, *237*, 871–881. [CrossRef]
10. Cascino, A.; Meli, E.; Rindi, A. A strategy for lightweight designing of a railway vehicle car body including composite material and dynamic structural optimization. *Railw. Eng. Sci.* **2023**, *31*, 340–350. [CrossRef]
11. Lang, D.; Radford, D.W. Design Optimization of a Composite Rail Vehicle Anchor Bracket. *Urban Rail Transit* **2021**, *7*, 84–100. [CrossRef]
12. Kuczek, T.; Szachniewicz, B. Topology optimisation of railcar composite structure. *Int. J. Heavy Veh. Syst.* **2015**, *22*, 375. [CrossRef]
13. Šťastniak, P.; Šalantay, P.; Harušinec, J.; Moravčík, M.; Suchánek, A. Application of topological optimization methods in the development of new generation freight railway wagons. *AIP Conf. Proc.* **2023**, *2976*, 030020. [CrossRef]
14. Koenig, J. Integral consideration of the lightweight design for railway vehicles. In *Young Researchers Seminar*; Technical University of Denmark: Lyngby, Denmark, 2011.
15. Cho, J.G.; Koo, J.S.; Jung, H.S. A lightweight design approach for an EMU carbody using a material selection method and size optimization. *J. Mech. Sci. Technol.* **2016**, *30*, 673–681. [CrossRef]
16. Miao, B.; Luo, Y.; Peng, Q.; Qiu, Y.; Chen, H.; Yang, Z. Multidisciplinary design optimization of lightweight carbody for fatigue assessment. *Mater. Des.* **2020**, *194*, 108910. [CrossRef]
17. Yin, J.; Zhang, Q.; Li, X.; Zhu, Y.; Liu, Z.; Liu, Y.; Sha, Z. Research on Lightweight Rail Vehicle Body Based on Sensitivity Analysis. *J. Eng. Technol. Sci.* **2024**, *56*, 353–366. [CrossRef]
18. Srivastava, P.K.; Shukla, S. Topology Optimization: Weight Reduction of Indian Railway Freight Bogie Side Frame. *Int. J. Mech. Eng.* **2021**, *6*, 4374–4383.
19. Harzheim, L.; Graf, G. A review of optimization of cast parts using topology optimization: I—Topology optimization without manufacturing constraints. *Struct. Multidiscip. Optim.* **2005**, *30*, 491–497. [CrossRef]
20. Yamamoto, M. Non-parametric optimization of railway wheel web shape based on fatigue design criteria. *Int. J. Fatigue* **2020**, *134*, 105463. [CrossRef]
21. Sharma, S.K.; Sharma, R.C.; Sharma, N. Combined multi-body-system and finite element analysis of a rail locomotive crashworthiness. *Int. J. Veh. Struct. Syst.* **2020**, *12*, 428–435. [CrossRef]
22. Guo, F.; Wu, S.C.; Liu, J.X.; Zhang, W.; Qin, Q.B.; Yao, Y. Fatigue life assessment of bogie frames in high-speed railway ve-hicles considering gear meshing. *Int. J. Fatigue* **2020**, *132*, 105353. [CrossRef]
23. Lucanin, V.J.; Simic, G.; Milkovic, D.D.; Cupric, N.U.; Golubovic, S.D. Calculated and experimental analysis of cause of the appearance of cracks in the running bogie frame of diesel multiple units of Serbian railways. *Eng. Fail. Anal.* **2010**, *17*, 236–248. [CrossRef]
24. Hou, K.; Kalousek, J.; Dong, R. A dynamic model for an asymmetrical vehicle/track system. *J. Sound Vib.* **2003**, *267*, 591–604. [CrossRef]
25. Song, Y.; Liu, Z.; Gao, S. Current Collection Quality of High-Speed Rail Pantograph-Catenary Considering Geometry Deviation at 400 km/h and Above. *IEEE Trans. Veh. Technol.* **2024**, *73*, 14415–14424. [CrossRef]
26. *EN 12663-1:2010+A2:2023*; Railway Applications—Structural Requirements of Railway Vehicle Bodies—Part 1: Locomotives and Passenger Rolling Stock (and Alternative Method for Freight Wagons). Available online: https://www.en-standard.eu/bs-en-12663-1-2010-a2-2023-railway-applications-structural-requirements-of-railway-vehicle-bodies-locomotives-and-passenger-rolling-stock-and-alternative-method-for-freight-wagons/?srsltid=AfmBOoqIIoj8nfUYWdVX3a3zh6pMAK3CUpoKIhb7tzcrTfkeBwwVkx76 (accessed on 24 October 2024).

27. Cascino, A.; Meli, E.; Rindi, A. A New Strategy for Railway Bogie Frame Designing Combining Structural–Topological Optimization and Sensitivity Analysis. *Vehicles* **2024**, *6*, 651–665. [CrossRef]
28. *EN 13749:2021+A1:2023*; Railway Applications—Wheelsets and Bogies—Method of Specifying the Structural Requirements of Bogie Frames. Available online: https://www.en-standard.eu/bs-en-13749-2021-a1-2023-railway-applications-wheelsets-and-bogies-method-of-specifying-the-structural-requirements-of-bogie-frames/ (accessed on 24 October 2024).
29. Jameson, A. Gradient based optimization methods. In *MAE Technical Report No. 2057*; Princeton University: Princeton, NJ, USA, 1995.

MDPI AG
Grosspeteranlage 5
4052 Basel
Switzerland
Tel.: +41 61 683 77 34

Vehicles Editorial Office
E-mail: vehicles@mdpi.com
www.mdpi.com/journal/vehicles